CURRICULUM AND EVALUATION

STANDARDS

FOR SCHOOL MATHEMATICS

ADDENDA SERIES, GRADES 9–12

# DATA ANALYSIS AND STATISTICS ACROSS THE CURRICULUM

Gail Burrill
John C. Burrill
Pamela Coffield
Gretchen Davis
Jan de Lange
Diann Resnick
Murray Siegel

Consultants

Harold L. Schoen
Daniel J. Teague

Christian R. Hirsch, Series Editor



NCTM
NATIONAL COUNCIL OF
TEACHERS OF MATHEMATICS

Library of Congress Cataloging-in-Publication Data:

Data analysis and statistics across the curriculum / Gail Burrill ... [et al.] ; consultants, Harold L. Schoen, Daniel J. Teague.
p. cm. — (Curriculum and evaluation standards for school mathematics addenda series. Grades 9–12)
Includes bibliographical references.
ISBN 0-87353-329-1
I. Burrill, Gail. II. Series.
QA276.18.D37 1992
001.4'22'071273—dc20 92-16923
CIP

The publications of the National Council of Teachers of Mathematics present a variety of viewpoints. The views expressed or implied in this publication, unless otherwise noted, should not be interpreted as official positions of the Council.

Printed in the United States of America

# TABLE OF CONTENTS

## *FOREWORD*

As the *Curriculum and Evaluation Standards for School Mathematics* (NCTM 1989) was being developed, it became apparent that supporting publications would be needed to aid in interpreting and implementing the curriculum and evaluation standards and the underlying instructional themes. A Task Force on the Addenda to the *Curriculum and Evaluation Standards for School Mathematics,* chaired by Thomas Rowan and composed of Joan Duea, Christian Hirsch, Marie Jernigan, and Richard Lodholz, was appointed by Shirley Frye, then NCTM president, in the spring of 1988. The Task Force's recommendations on the scope and nature of the supporting publications were submitted in the fall of 1988 to the Educational Materials Committee, which subsequently framed the Addenda Project for NCTM Board approval.

Central to the Addenda Project was the formation of three writing teams to prepare a series of publications targeted at mathematics education in grades K–6, 5–8, and 9–12. The writing teams consisted of classroom teachers, mathematics supervisors, and university mathematics educators. The purpose of the series was to clarify the recommendations of selected standards and to illustrate how the standards could realistically be implemented in K–12 classrooms in North America.

The themes of problem solving, reasoning, communication, and connections have been woven throughout each volume in the series. The use of technological tools and the view of assessment as a means of guiding instruction are integral to the publications. The materials have been field tested by teachers to ensure that they reflect the realities of today's classrooms and to make them "teacher friendly."

We envision the Addenda Series being used as a resource by individuals as they begin to implement the recommendations of the *Curriculum and Evaluation Standards.* In addition, volumes in a particular series would be appropriate for in-service programs and for preservice courses in teacher education programs.

On behalf of the National Council of Teacher of Mathematics, I would like to express sincerest appreciation to the authors, consultants, and editor who gave willingly of their time, effort, and expertise in developing these exemplary materials. Special thanks are due to Lois Edwards, Pat Hopfensperger, James Landwehr, Richard Scheaffer, Ken Sherrick, Ken Travers, and Kenneth Vos who provided the authors insights and useful ideas on teaching statistics and who reviewed drafts of the material as this volume progressed. Finally, the continuing technical assistance of Cynthia Rosso and the able production staff in Reston is gratefully acknowledged.

Bonnie H. Litwiller
Addenda Project Coordinator

## *PREFACE*

The *Curriculum and Evaluation Standards for School Mathematics,* released in March 1989 by the National Council of Teachers of Mathematics, has focused national attention on a new set of goals and expectations for school mathematics. This visionary document provides a broad framework for what the mathematics curriculum in grades K–12 should include in terms of content priority and emphasis. It suggests not only what students should learn but also how that learning should occur and be evaluated.

Although the *Curriculum and Evaluation Standards* specifies the key components of a quality contemporary school mathematics program, it encourages local initiatives in realizing the vision set forth. In so doing, it offers school districts, mathematics departments, and classroom teachers new opportunities and challenges. The purpose of this volume, and others in the Addenda Series, is to provide instructional ideas and materials that will support implementation of the *Curriculum and Evaluation Standards* in local settings. It addresses in a very practical way the content, pedagogy, and pupil assessment dimensions of reshaping school mathematics.

### *Reshaping Content*

The curriculum standards for grades 9–12 identify a common core of mathematical topics that *all* students should have the opportunity to learn. The need to prepare students for the workplace, for college, and for citizenship is reflected in a broad mathematical sciences curriculum. The traditional strands of algebra, functions, geometry, and trigonometry are balanced with topics from data analysis and statistics, probability, and discrete mathematics.

The increased attention given to data analysis and statistics is accompanied by a shift in perspective on how topics in this strand are treated. The standard on statistics precedes the probability standard to signal a call for a move away from the usual treatment of combinatorial counting problems as a precursor to the study of statistics. The standard, itself, emphasizes the importance of developing statistical thinking, not simply learning statistical procedures. In particular, if statistics is "making sense out of data," then the curriculum should provide numerous experiences involving students in designing experiments, collecting and organizing data, representing the data with visual displays and summary statistics, analyzing the data, communicating the results, and (where appropriate) searching for prediction models. Central to the first phase is understanding sampling; central to the last is understanding confidence intervals. Throughout, there should be an emphasis on developing a critical attitude toward the uses of statistics.

*Data Analysis and Statistics across the Curriculum* links the content proposed in the *Curriculum and Evaluation Standards* to that of current programs. Ways for integrating data analysis and statistics with the algebra, functions, and geometry strands are described and illustrated with numerous examples. The importance of communication in statistics is stressed. The possibilities statistics offers for connecting mathematics with the students' world and with other school subjects is a recurring theme.

A special *Try This* feature appearing throughout the volume provides exercises, problems, and explorations for use with students. We hope that these will pique your interest and that you will use them productively.

More extensive investigations and problems appear as blackline masters at the end of each chapter. These activities are appropriate for students at varied levels. Solutions for, and comments on, these activities appear in the Appendix. Chapter 7 includes a rich and extensive collection of questions for project work. Additional sources of ideas and materials supportive of the data analysis and statistics strand are identified in a selective annotated bibliography at the end of the volume.

***Reshaping Pedagogy***

The *Curriculum and Evaluation Standards* paints mathematics as an activity and a process, not simply as a body of content to be mastered. Throughout, there is an emphasis on doing mathematics, recognizing connections, and valuing the enterprise. Hence, standards are presented for Mathematics as Problem Solving, Mathematics as Communication, Mathematics as Reasoning, and Mathematical Connections. The intent of these four standards is to frame a curriculum that ensures the development of broad mathematical power in addition to technical competence; that cultivates students' abilities to explore, conjecture, reason logically, formulate and solve problems, and communicate mathematically; and that fosters the development of self-confidence.

Realization of these process and affective goals will require, in many cases, new teaching-learning environments. The traditional view of the teacher as authority figure and dispenser of information must give way to that of the teacher as catalyst and facilitator of learning. To this end, the standards for grades 9–12 call for increased attention to—

- actively involving students in constructing and applying mathematical ideas;
- using problem solving as a means as well as a goal of instruction;
- promoting student interaction through the use of effective questioning techniques;
- using a variety of instructional formats—small cooperative groups, individual explorations, whole-class instruction, and projects;
- using calculators and computers as tools for learning and doing mathematics.

*Data Analysis and Statistics across the Curriculum* reflects the new methodologies supporting new curricular goals. The classroom-ready activity sheets at the end of the chapters provide tasks and problem situations that require students to experiment; to collect, represent, analyze, and fit curves to data; and to evaluate and communicate the results of these processes. These activities are ideally suited to cooperative group work. As with all student investigations, it is important that provisions be made for students to share their experiences, clarify their thinking, generalize their discoveries, and provide convincing explanations. The previously mentioned *Try This* feature appearing throughout the volume furnishes more structured tasks offering opportunities for mathematical modeling, problem solving, reasoning in context, and lively classroom discourse.

*Teaching Matters* is another special feature of this book. These captioned margin notes supply helpful instructional suggestions, including ideas on motivation and on the effective use of technological tools. They also identify possible student misconceptions and some difficulties students might encounter with certain topics and suggest how these can be anticipated and addressed in instruction.

At the heart of changing patterns of instruction are the growing potentialities of technology. The standards for grades 9–12 assume that students will have access to graphing calculators and that computers will be available for demonstration purposes as well as for individual and group work. As illustrated in chapters 1 and 3–5, computer software with the capability to store, manipulate, and display data sets and summary statistics permits students to concentrate on interpretation and decision making. Many of these software packages also have curve-fitting routines that provide the equation and graph of the best fitting curve as well as a correlation coefficient and residuals. Many of these capabilities are now also available on graphing calculators.

Both computers and calculators have random number generators that are useful in simulations as exemplified in chapters 1 and 6. Monte Carlo methods supported by these tools can provide more students greater access to important ideas involving probability. The annotated bibliography provides additional information on computer software tools and on instructional materials for use with graphing calculators.

***Reshaping Assessment***

Complete pictures of classrooms in which the *Curriculum and Evaluation Standards* is being implemented not only show changes in mathematical content and instructional practice but also reflect changes in the purpose and methods of student assessment. Classrooms where students are expected to be engaged in mathematical thinking and in constructing and reorganizing their own knowledge require adaptive teaching informed by observing and listening to students at work. Thus informal, performance-based assessment methods are essential to the new vision of school mathematics.

Analysis of students' written work remains important. However, single-answer paper-and-pencil tests are often inadequate to assess the development of students' abilities to analyze and solve problems, make connections, reason mathematically, and communicate mathematically. Potentially richer sources of information include student-produced analyses of problem situations, solutions to problems, reports of investigations, and journal entries. Moreover, if calculator and computer technologies are now to be accepted as part of the environment in which students learn and do mathematics, these tools should also be available to students in most assessment situations.

*Data Analysis and Statistics across the Curriculum* reflects the multidimensional aspects of student assessment and the fact that it is integral to instruction. *Assessment Matters,* a special margin feature, provides suggestions for assessment techniques as well as ideas for test items related to the content under discussion. Possible ways of evaluating student projects are illustrated in chapter 7 with samples of actual student work. The volume concludes with a chapter devoted to methods for assessing statistical understanding. Included is discussion of two promising new ideas, the two-stage test and the student-made test.

***Conclusion***

The standards on statistics and probability for grades 9–12, informed by research and based on the wisdom of practice, represent the consensus of the profession as to the central position that data and chance ought to occupy in the high school curriculum. Data and questions related to representation, analysis, and prediction are at the heart of statistics. They also provide a context for the motivation and application of important ideas in both algebra and geometry. Seizing opportunities to blend new and traditional topics holds promise as a strategy for initiating change while avoiding the problem of "add-ons" to the curriculum.

Sustainable change must occur first in the hearts, minds, and classrooms of teachers and then in their departments and school districts. Individually we can initiate the process of change; collectively we can make the vision of the *Curriculum and Evaluation Standards* a reality. We hope you will find the information in this book valuable in translating the vision of the *Standards* into practice.

Christian R. Hirsch, Editor
Grades 9–12 Addenda Series

## CHAPTER 1
## STATISTICS IN THE CURRICULUM

How can we teach statistics when we also have to teach algebra and geometry and trigonometry and calculus? What can we leave out to make room? Where can we "put" some probability or statistics that will not interfere with real mathematics? Questions like these are heard in classrooms across the nation as teachers begin to think about implementing the *Curriculum and Evaluation Standards for School Mathematics* (NCTM 1989). As the traditional curriculum is examined, all too often the essence of the *Standards* is reduced to a checklist with a few old topics deleted and new ones added. Out with radicals, in with box plots! The *Standards*, however, is not a list of objectives. It is a framework for transforming the mathematics curriculum to better prepare *all* students to be mathematically literate.

Two essential questions to consider when discussing curriculum are, "What do students need to know outside the classroom?" and "How will they have to communicate what they know?" On the job, in the market, and in the voting booth, are they more likely to have to make decisions that require factoring a polynomial or that deal with issues of probability and uncertainty? The disposal of nuclear waste, acid rain, the ozone layer, AIDS—all must be resolved by tomorrow's citizens. Associated with each of these issues are questions of enormous consequence involving small probabilities. When these questions are decided, people must be able to express themselves in oral and written arguments and be able to understand and interpret the arguments of others. They must be able to speak about mathematical notions and in the language of mathematics, statistics, and algebra.

We live in an information age. Because statistics is the science of collecting and analyzing data and turning data into information, statistics is as important as—and for many students, more important than—traditional mathematics. However, teaching statistics does not preclude teaching "real" mathematics. In fact, statistics can provide motivation for studying traditional mathematics topics that remain fundamental. For example, certain questions—How can you determine the chances of winning a ball game? Can you predict how tall people will be if you know their height at age five? Do tall people run faster? Which brand of batteries will last the longest in a CD player?—call for the collection of data. In the process of analyzing the data, students may discover the need to study logarithms or use vectors to find the best model for their data. In order for statistics to be used in this manner, it is essential that statistical ideas be introduced early in the high school curriculum and then systematically developed across the curriculum so that statistical thinking becomes a habit.

### STATISTICS FOR SCHOOLS

What should the classroom teacher try to accomplish when "teaching statistics"? Just what is statistics? Statistics is "making sense of data." In the world outside the classroom, statistics is used primarily as a tool to solve a problem or answer a question. Thus, the teaching of statistics should begin with some problem or question or concern: How much spending money do high school students have? Is there any relation between the time spent on homework and test grades? The investigation of a problem begins with data. If the data are not already available, some scheme must be designed for collecting the data. The first step is to

*Try This:* ***Have students write their name three times in succession. All will be different. There is natural variability in writing your name. Now have pairs of students exchange papers and write each other's name. Distribute the papers around the class and ask students to identify the imposter. One of the principal aspects of the study of statistics is in trying to determine not just when things vary (they always vary) but when they vary enough to be called different.***

*Assessment Matters:* ***Expect students to find using statistics to describe a set of data much easier than using statistics to make inferences. In particular, inferential statistics requires an understanding of the distinction between an empirical probability and a theoretical probability, a distinction that underlies the concept of a probability distribution. Many successful graduate students never fully grasp this concept.***

make sure that the data collected are relevant to the problem. The next step is to summarize the data, either graphically or numerically, and then look for possible solutions suggested by the summary.

A key step is to recognize that everything is subject to variability. A fair coin tossed 100 times may land heads up 45 or 60 times; two measurements of the same object may differ by 0.2 mm. If the amount of variability is very small in relation to the question of interest, the solution to the problem may be obvious. If random samples of trips in car A repeatedly give 18 to 22 miles per gallon and random samples of trips in car B repeatedly give 33 to 37 miles per gallon, there is no need to continue investigating the question of whether there is a difference in fuel consumption between the two makes of cars.

Suppose, however, what you want to learn is not obvious. If battery A has an average life span of 42 hours and battery B has an average life span of 39 hours (based on samples of only a few batteries each), has a difference in battery life really been established? Is a three-hour difference large enough to be meaningful? Is a 2 percent gain in test scores significant or is it merely random variation? To answer questions such as these, statisticians create a mathematical model, whose behavior can be predicted, to describe the problem. This way of thinking, called *statistical inference*, uses the concept of randomness in two different ways.

In the first way, the behavior of a population is assumed, and the behavior of a random sample selected from this population is compared to this known or assumed standard. The question then becomes How does the sample measure up? For example, the average death rate due to a certain kind of cancer is known for the general population. Consider the subset of people who live near a toxic dump. Is their death rate abnormally high when compared to the death rate expected from a randomly selected sample of the population?

In the second form of inference, the behavior of a sample is known, and the objective is to infer from that sample some knowledge about the behavior of the population. A sample of a given make of car averages 25 miles per gallon. What does this tell us about the mileage rate for all cars of this make? It is at this stage of statistical inference that probability enters the picture. The sampling must be done randomly, that is, every member of the population must have an equal chance of being selected, and the selection of one member must be independent of the selection of another. The comparison or decision is usually quantified in probabilistic language: a 20 percent chance or 90 percent confident. The concept of probability necessary for reasoning statistically at the high school level is not complicated and can be based almost entirely on the notion of counting the number of times an event occurs with many repetitions of the experiment. Simulation can be a powerful tool in statistical inference. Observing random behavior by simulating an event and recording the results can avoid the use of complicated mathematical models and foster understanding of the situation.

Statistics, then, plays two major roles. The first is data exploration, sometimes with little regard for how the data were collected. Questions of interest are What do the data mean? What information is in the data? and What interesting questions deserve further investigation? The second role is that of statistical inference, which requires careful plans for data collection involving randomness so that patterns can be predicted and bias reduced. The focus is either to verify an assumed model (compare a sample to a standard) or to build a model (describe a population from a sample). Much of the work at the school level deals with the first role because of the environment in which students live, but they should also understand the basic principles of the second.

♦ ♦ ♦ ♦ ♦ ♦ ♦ ♦

### *STATISTICS IN THE CLASSROOM*

What does all this mean to the classroom teacher? The principal elements in the description of statistics are the problem, data, summarization of data, variation, and the relationship of samples to population. The focus of the curriculum and teaching should be on those elements. Consider, for example, a recent episode in a ninth-grade algebra class studying linear equations. Students raised questions such as "Why are lines important?" and "What good is it to graph information and try to make lines?" as they attempted to organize what they had been doing for the last several days. The answer came in the form of another question: "Do tall people run faster?" Students immediately took sides. One student mentioned the tall half-mile conference champion on the girl's track team, and someone else countered with the running back on the football team who is short and fast. After some prodding, the students decided to set up an experiment to investigate the problem. They split into groups of four; one group laid out a course on the track, another brought in five stop watches, and another prepared a data sheet on which to record each student's height and the time it took to run the length of the course. The next day, attired in tennis shoes and jeans and proceeding with a minimal amount of confusion, students collected the data and returned to the classroom. Armed with data sheets, graph paper, and some questions to guide their analysis, students organized the data, made scatterplots, and looked for a pattern in their data that would help them to answer the question. See figure 1.1.

***Teaching Matters:* Unlike most traditional classroom mathematics, there is no single correct numerical answer to a problem such as "Do Tall People Run Faster?" Instead, each possible solution depends heavily on the assumptions and perception of the problem solver. This dramatically increases the need for communication skills on the part of the student because a teacher cannot assume evidence that is not presented.**

***Try This:* Set up an experiment to answer the question "Is there any relation between height and weight?"**

| Name | Height (in.) | Time (sec.) |
|---|---|---|
| 1. Amy | 66 | 8.44 |
| 2. Jim | 66 | 7.00 |
| 3. Tony | 71 | 7.54 |
| 4. Corry | 68 | 8.91 |
| 5. Joe | 69 | 7.44 |
| 6. Matt | 62 | 8.10 |
| 7. Teresa | 61 | 8.40 |
| 8. Jenny | 69 | 8.00 |
| 9. Renee | 64 | 9.21 |
| 10. Todd | 66 | 7.72 |
| 11. Scott | 72 | 7.92 |
| 12. Marty | 67 | 7.78 |
| 13. Angy | 68 | 9.00 |
| 14. Dan | 69 | 8.44 |
| 15. Mike | 71 | 7.74 |
| 16. Stephanie | 65 | 8.45 |
| 17. Rachel | 63 | 8.55 |
| 18. Eric H. | 69 | 8.20 |
| 19. Nicole | 68 | 8.55 |
| 20. David | 67 | 8.50 |
| 21. Jason | 67 | 7.00 |
| 22. Vaishalee | 61 | 9.16 |
| 23. Mrs. B | 66 | 13.88 |
| 24. Evan | 68 | 8.10 |
| 25. Eric V. | 68 | 7.66 |
| 26. Nick | 67 | 8.40 |
| 27. Sheila | 60 | 9.48 |

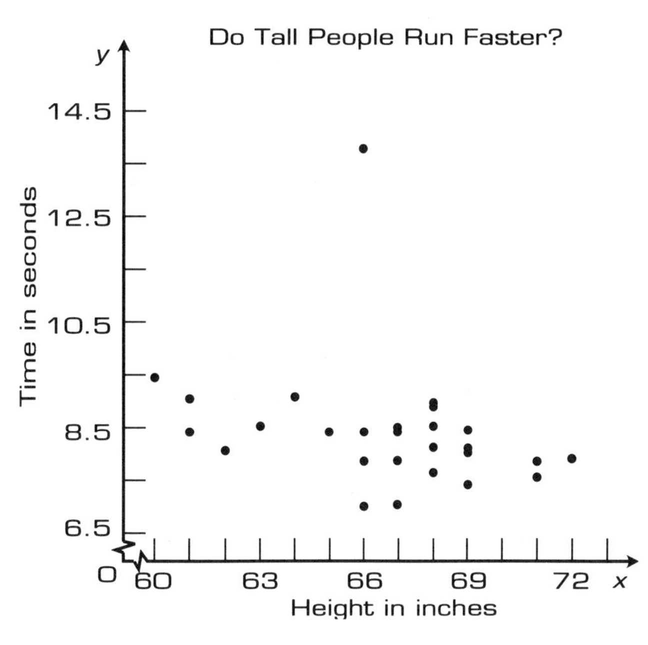


***Fig. 1.1***

*Teaching Matters:* ***Encourage students to write up the conclusions they draw from experiments. For procedural tips, check with a science teacher on lab-report formats.***

*Teaching Matters:* ***The linear association between speed and height is not strong, although there is some evidence of a trend. The students should recognize this as they begin to make conclusions. Older students can find the correlation coefficient with and without the outlier and discuss the difference.***

Heated discussion followed. Jason and Joe finished at the same time, yet Joe's time was 0.44 seconds faster than Jason's time (a very important difference to ninth-grade boys); the timers were all messed up. It was not really fair to use the time from only one trial; an average of several trials would have been much better. The course was too short; the time would be different if it had been longer. Some people did not run as fast as they could have; they messed up the results. Some people had better shoes for running; they were faster. One person (the teacher) was too old to compare to the rest!

A student summary of the results is given below.

> Do Tall People Run Faster?
>
> The results of our experiment showed that there is some relation between how fast people run and how tall they are. The correlation is negative; the shorter the person, the longer it takes them to run the course. The equation of the median fit line was $y = -.076x + 13.39$, and the slope of the line indicates the rate of change of time in terms of change in height. A person takes about .08 seconds off their time for every inch taller they are. Keith had a sore knee and couldn't run, so we used the equation to predict Keith's time. He was 66 inches tall, so his time would have been about 8.37 seconds, but this would only be an estimate. It probably would be within 1.00 second either way.
>
> Our outlier did not really affect our results because we used medians, and outliers don't have much effect on them. We did have some trouble collecting the data because the timers weren't accurate and also because some people really didn't try to run very fast. When we made the line, we had lots of points in a row and they were hard to separate, so we tossed coins to see which part to put the points in. This made our answers all different but not by much.

Chaos in the classroom? Not at all. The objections and concerns were valid and needed to be factored into any conclusion. Isn't that what making decisions in the world outside of the classroom is all about? Are most decisions clear-cut, uncluttered by information? Sadly enough, most of the mathematics in textbooks is not constrained by context. The mathematics most students will do in life, however, begins with a context, and mathematics must be interpreted in terms of that context. Students need opportunities in school to experience the mathematics they will eventually practice outside of school. What did the algebra students learn about statistics? They learned the process, the use of data, the impact of variability, the importance of organizing their results, and the importance of a context in making conclusions. What did the students learn about mathematics? They learned why lines and equations are important and how they are used. They learned that planning the experiment carefully before doing it saves many arguments afterward. The first activity at the end of this chapter can be used to introduce this experiment in other classrooms.

*Teaching Matters:* ***In every class, assign small projects that involve students in collecting and analyzing their own data.***

*Assessment Matters:* ***Situations from magazines and newspapers where real data are used can also provide settings for assessment. Have students summarize these data in ways they have learned, respond to questions about the data summary, and discuss how the data may or may not be used to make certain inferences.***

### INCLUDING STATISTICS IN THE CURRICULUM

One way to incorporate statistics in the curriculum is to begin both algebra and advanced algebra with a two-to-three-week unit, then bring in additional ideas throughout the year as time and topic are appropriate. In algebra, start by using graphs and charts from newspapers and magazines on subjects of interest to the students to help them understand why statistics is important and to begin interpreting information presented in articles and graphs. In an initial statistics unit, students should learn to be critical of data and to organize data using simple graphical techniques such as histograms, line plots, stem-and-leaf plots, box plots, and scatterplots.

◆ ◆ ◆ ◆ ◆ ◆ ◆ ◆

For example, consider the data in table 1.1. Students could be asked to organize the data with a stem-and-leaf plot and use the plot to write a paragraph about the speeds of animals. Most well-trained mathematics students immediately begin by constructing the graph and do not stop to reflect on what the numbers mean. Information expressed numerically is usually not challenged! How fast is 56 kilometers per hour? What does it mean for a human to run 43 kilometers per hour? After some prodding and a few calculations, students recognize that an important fact must be missing in the information about the data. The numbers must be sprint or maximum speeds, a fact not obvious because the units are metric and unfamiliar to most students. Understanding what the data represent should be a first step, not the last; students need to be taught to approach data critically.

*Teaching Matters:* ***Students have to be directed about what to write and how to write when interpreting data such as in table 1.1. They tend to say, "Some of the animals went slower than the rest" or "Lots of animals are fast" without quantifying their statements and without referring to the data. They also tend to describe the obvious and have to be taught to look for patterns. For tips on how to teach students to write, talk to an English teacher in your school.***

Table 1.1
Speed of Animals (in km/h)

| | | | |
|---|---|---|---|
| antelope | 98 | giraffe | 51 |
| bear | 48 | greyhound | 62 |
| cat | 48 | horse | 76 |
| cheetah | 110 | human | 43 |
| chicken | 14 | lion | 80 |
| coyote | 68 | pig | 18 |
| deer | 56 | rabbit | 56 |
| elephant | 40 | turkey | 24 |
| elk | 72 | warthog | 49 |
| fox | 51 | zebra | 64 |
| jackal | 56 | | |

Source: *World Almanac Book of Facts,* 1990 edition © Newspaper Enterprise Association, Inc. 1989, New York, NY

Other important topics for a beginning unit include numerical summaries for the center and variation of data. Students should learn how to handle both univariate (one variable) and bivariate (two variable) data and be able to compare two data sets using both graphical and numerical measures (table 1.2).

Table 1.2.
Typical Ways to Summarize Data

| | One variable | Two variables |
|---|---|---|
| Center | mean, median, mode | equation |
| variation | standard deviation | mean squared error |
| | interquartile range | correlation |
| | quartiles | |
| | range | |
| | outliers | |

An understanding of sampling as well as the ability to recognize when samples may be biased is important. The focus should be on exploring data by applying several techniques to the same set of data and using the information given in the graphs or numerical summaries to solve a problem. Two sources for this unit are *Exploring Data* (Landwehr and Watkins 1986) from the Quantitative Literacy series and *Data Visualization* (Lange and Verhage 1989). Students should come to recognize that in statistics there is often no single correct numerical answer, but instead there are observations that can be made about the data. A

*Try This:* ***Have students explore whether a family is more likely to have 6 girls in a row or to have children in an alternating sequence, such as girl-boy-girl-boy-girl-boy.***

◆ ◆ ◆ ◆ ◆ ◆ ◆ ◆

*Assessment Matters:* ***There are many common misconceptions about probability. (See, for example, the 1981 NCTM Yearbook,* Teaching Statistics and Probability.*) Being on the lookout for some of these misconceptions will improve a teacher's skill at diagnosing students' difficulties.***

*Try This:* ***Draw the figure below on a sheet of paper. Use the copy machine to reduce and enlarge the figure. Have students construct the boxes and fill them level with salt. Compare the weights of the filled boxes with the amount of reduction induced by the photocopying. What function describes the relationship?***

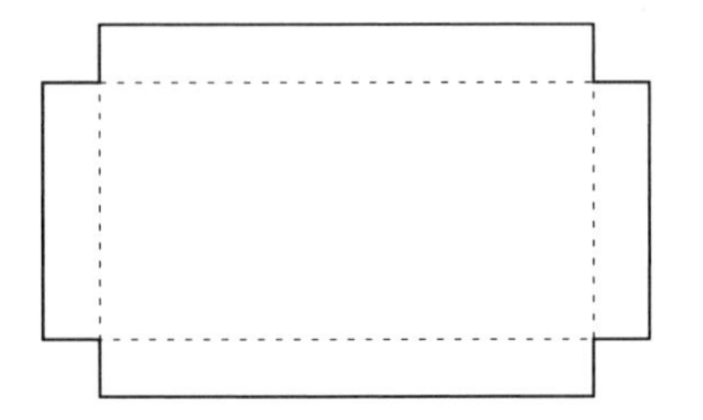

*Try This:* ***What is the sum of the exterior angles of a pentagon? Have students collect measurements from a sample of randomly selected pentagons, record their results in a stem-and-leaf plot, and try to make conclusions from observing the distribution.***

*Try This:* ***Is there any relation between the sides of a 30°-60° right triangle? Have students measure the hypotenuse and longest leg of samples of 30°-60° right triangles. Plot (leg, hypotenuse) and analyze the graph to determine if there is a relation. If so, write an equation to describe it.***

variety of approaches are generally reasonable, and answers are usually expressed as written sentences or paragraphs. Once this foundation has been established, statistical topics can be included easily and naturally at other points in the curriculum.

Connections among mathematical topics can give students a powerful set of tools to solve problems and can enhance their appreciation of the consistency and beauty of mathematics. Examining content areas from different perspectives is an effective way of building these connections. For example, when discussing reasoning and proof in geometry, explore simulation as a process to support or verify a conclusion. In second-year algebra, begin the year with graphing and introduce the correlation coefficient and the interpretation of residual plots. In precalculus encourage the study of compositions and inverses as well as logarithmic and exponential functions by using graphing calculators and computer graphics to re-express and linearize curvilinear data sets. Look for places in the curriculum where real data can be used, where students can begin with a question and generate data to explore possible answers, where variation is a reality. Summarize test data by using means and standard deviations or interquartile ranges or equations and discuss what factors to consider when judging the reliability of each of these summary measures. Collect assignments on a random basis by giving each student a number and having an assignment lottery each day. When the same number is called three days in a row and students claim the lottery is fixed, seize the opportunity to discuss randomness.

When studying percent, consider the difference between the following two problems:

1. 30% of the students in the school are left-handed. If there are 25 students in our class, how many left-handed students should we expect to find?
2. 12 of the 25 students in our class are left-handed. What percent of the students in school will be left-handed?

Problem 1 represents the first type of statistical inference, using probability and randomness to predict properties of a sample from a population standard. The second problem illustrates the alternative process, making some statements about an unknown population on the basis of knowledge from a sample. Both involve variation and prediction in real situations and are not simple computational exercises. Reasonable answers can easily be obtained by simulation (Burrill 1990).

Statistical ideas can also be incorporated in the geometry strand. The activity "What Is the Probability That Spaghetti Pieces Form a Triangle?" at the end of this chapter connects ideas of chance and data with the triangle inequality. In the study of area have students measure the sides and area of a variety of squares, plot the results, and develop the quadratic model $y = x^2$ to describe the relation between the two variables. Is there any variability in the pattern? What might cause this? What would the plot of the edge versus the volume of a cube look like? Extend the study of area by using area models for probability. Interesting examples can be found in *Probability* (Lappan et al. 1986), from the Michigan Middle School Mathematics Project, and in *Geometric Probability* (North Carolina School of Science and Mathematics 1989).

In the study of similarity, consider whether any two normal distributions are similar. Create a distribution by taking the heights for children of the same grade of a large sample of girls and another by using the heights of a large sample of boys. The smoothed distributions should approximate a bell-shaped curve, close to a normal distribution. Are the two bell-shaped distributions similar? Why or why not? Estimate the middle two-thirds of the area (the area within 1 standard deviation of the mean) for each distribution either by counting unit squares or by finding the percentage of scores contained in the interval. Are the areas approximately equal?

In coordinate graphing, use real data to make scatterplots; have students graph characteristics of the class—for example, hours of TV watched per week versus hours of homework; the number of children in the family versus the number of phones; the age of the mother versus the age of the father. Use the line $y = x$ to analyze inequalities and compare two sets of variables. Have students analyze plots of data for patterns by graphing ordered pairs by hand, with the computer, and on hand-held graphing calculators. As students become more sophisticated mathematically, explore best-fit lines, correlation, and information provided by an analysis of residuals. Incorporate calculators and computer software in investigating data, exploring models, and generating random numbers.

Teaching mathematics using data is a mind-set, and because it is foreign to the way mathematics is currently taught, it will not be easy. The teacher must begin to think, "What problem can be studied in this context? How can real data be used in this lesson? What kind of information can students generate that will enhance a given mathematical objective?" The vision of the *Curriculum and Evaluation Standards* is not "out with the old" but a change in perspective about what is important today. The following classroom-ready activities illustrate this shift in perspective.

Students should be informed a day before conducting Activity 1 so that they wear appropriate clothing and shoes. Five or six stop watches will be needed. A course of approximately 50 yards will need to be laid out. On the day of the experiment, measure the height of each student, and then have students run in sets of five with a timer at the finish mark of each lane. Heights and times for each runner can be recorded and plotted on copies of sheet 1. Graphing calculators or computer software such as Data Insights also can be used to plot the data and fit a line.

*Try This:* ***What shape would you expect to get if you plotted the radius and circumference for a sample of circles? What would you expect for the slope? Measure a sample, plot the results, and use the plot to answer the questions.***

*Try This:* ***Gather a number of jar lids of different sizes. Have the students put dried peas into the lids to fill the bottom of the lid. No pea should lie on top of another. What is the relationship between the radius of the lid and the number of peas that fill the bottom? What is the relationship between the circumference of the lid and the number of peas that fill the bottom? Use a graphing calculator to compare the data with the curves $y = \pi x^2$ and $y = x^2/(2\pi)$.***

***ACTIVITY 1***
***DO TALL PEOPLE RUN FASTER?***

SHEET 1

1. Record the name, the height, and the running time for each student in your class.

| Name | Height (in.) | Time (sec.) |
|---|---|---|
| 1. | | |
| 2. | | |
| 3. | | |
| 4. | | |
| 5. | | |
| 6. | | |
| 7. | | |
| 8. | | |
| 9. | | |
| 10. | | |
| 11. | | |
| 12. | | |
| 13. | | |
| 14. | | |
| 15. | | |

| Name | Height (in.) | Time (sec.) |
|---|---|---|
| 16. | | |
| 17. | | |
| 18. | | |
| 19. | | |
| 20. | | |
| 21. | | |
| 22. | | |
| 23. | | |
| 24. | | |
| 25. | | |
| 26. | | |
| 27. | | |
| 28. | | |
| 29. | | |
| 30. | | |

2. Plot your results on the grid. Label the axes.

***ACTIVITY 1***
***DO TALL PEOPLE RUN FASTER? (CONTINUED)***

SHEET 2

3. Describe the experiment.

4. What factors had to be considered in setting up the experiment?

5. Summarize the results of the experiment. What conclusions can you make and why?

6. What variables might affect your results? Explain your answer.

7. Were there any outliers in the data? What might have caused this?

8. Predict how fast Keith, who is 66 inches tall, would run the course. How reliable is your prediction?

9. Make a stem-and-leaf plot of the times it took to run the course. Describe the typical time for the class. Where would you fit into the distribution?

**ACTIVITY 2** SHEET 1

## WHAT IS THE PROBABILITY THAT SPAGHETTI PIECES FORM A TRIANGLE?

1. Break a piece of spaghetti at two random points. Try to form a triangle with the three pieces. The pieces must touch end to end.
   a. Collect the class results. Number of triangles formed: ________ Number of instances of no triangle: ________
   b. Use these results to estimate the probability that a piece of spaghetti broken in this manner would form a triangle:

$$P(\text{triangle}) = \frac{\textit{number of triangles}}{\textit{number of initial pieces of spaghetti}} = \underline{\quad\quad}$$

2. Simulating the broken pieces of spaghetti can provide another way to estimate the probability. Generate two random numbers between 0 and 100. Use each as a side of a triangle. To find the third side, add the two random numbers and subtract their sum from 100. (100 represents the length of the spaghetti.) For example, if the random numbers were 25 and 39, the third side would be 100 – (25 + 39) = 36. To determine if the three numbers will form a triangle, cut off the ruler at the bottom of Sheet 2. Fold the ruler at point *A*, 25, the first random number. Fold it again at the point *B*, 64, the sum of the two random numbers. The third side will be 36 units. Will the ruler folded at those two points make a triangle?

   Working in pairs, generate 40 sets of random numbers using your calculator. For a TI-81, key in Math <Num><Int> Enter ( 99 * Math <Prb> Rand Enter ) + 1. The viewing screen will show Int(99*Rand) + 1. Press [Enter] to generate each random number. For the Casio fx-7000 or fx-8000, key in Int(99Ran#) + 1. Press the [EXE] key to generate each random number. (If you do not have a calculator that will generate random numbers, obtain a random number table from your teacher.) Record the lengths and results on the chart. Do not include a set of numbers if the sum is greater than 100.

| Trial | Random numbers | | Third side | Fold at | Triangle? |
|---|---|---|---|---|---|
| 1. | 25 | 39 | 36 | 64 | Yes |
| 2. | | | | | |
| 3. | | | | | |
| 4. | | | | | |
| 5. | | | | | |
| 6. | | | | | |
| 7. | | | | | |
| 8. | | | | | |
| 9. | | | | | |
| 10. | | | | | |
| 11. | | | | | |
| 12. | | | | | |
| 13. | | | | | |
| 14. | | | | | |
| 15. | | | | | |
| 16. | | | | | |
| 17. | | | | | |
| 18. | | | | | |
| 19. | | | | | |
| 20. | | | | | |

| Trial | Random numbers | | Third side | Fold at | Triangle? |
|---|---|---|---|---|---|
| 21. | | | | | |
| 22. | | | | | |
| 23. | | | | | |
| 24. | | | | | |
| 25. | | | | | |
| 26. | | | | | |
| 27. | | | | | |
| 28. | | | | | |
| 29. | | | | | |
| 30. | | | | | |
| 31. | | | | | |
| 32. | | | | | |
| 33. | | | | | |
| 34. | | | | | |
| 35. | | | | | |
| 36. | | | | | |
| 37. | | | | | |
| 38. | | | | | |
| 39. | | | | | |
| 40. | | | | | |

**ACTIVITY 2** SHEET 2

## WHAT IS THE PROBABILITY THAT SPAGHETTI PIECES FORM A TRIANGLE? (CONTINUED)

3. Use the results from the chart to estimate the probability of forming a triangle. ______________

4. How do these results compare to your first estimate? ______________

   What might account for any differences in the estimates? ______________

5. Under what condition will the lengths generated form a line segment? ______________

6. Look carefully at the set of lengths that form triangles. What conjecture can be made about the relation among the lengths of the sides of a triangle? ______________

   The probability that the pieces of spaghetti will form a triangle can also be found using algebra and geometry. Let the lengths of the sides generated be $x$ and $y$. The length of the third side will then be $100 - (x + y)$.

$x$ $y$ $100 - (x + y)$

7. Why must $x + y < 100$? ______________

   Thus, the domain for the problem is $y < 100 - x$, $x > 0$, and $y > 0$.

8. Use your conjecture from exercise 6 to explain why the following must also be true:

   a. $x + y > 100 - (x + y)$ ______________

   b. $x + 100 - (x + y) > y$ ______________

   c. $y + 100 - (x + y) > x$ ______________

9. Simplify each of the three inequalities in part 8.

   a. ______________

   b. ______________

   c. ______________

10. To find the solution to the system, graph the inequalities on the grid or use a graphing calculator.

11. Shade and then calculate the area of the region satisfying the system of inequalities.

12. Find the probability that the three lengths will form a triangle.

$$P(\text{triangle}) = \frac{\textit{area of region representing the solution to the system of inequalities}}{\textit{area of region representing the domain}}$$

$= \_\_\_\_\_\_\_\_$

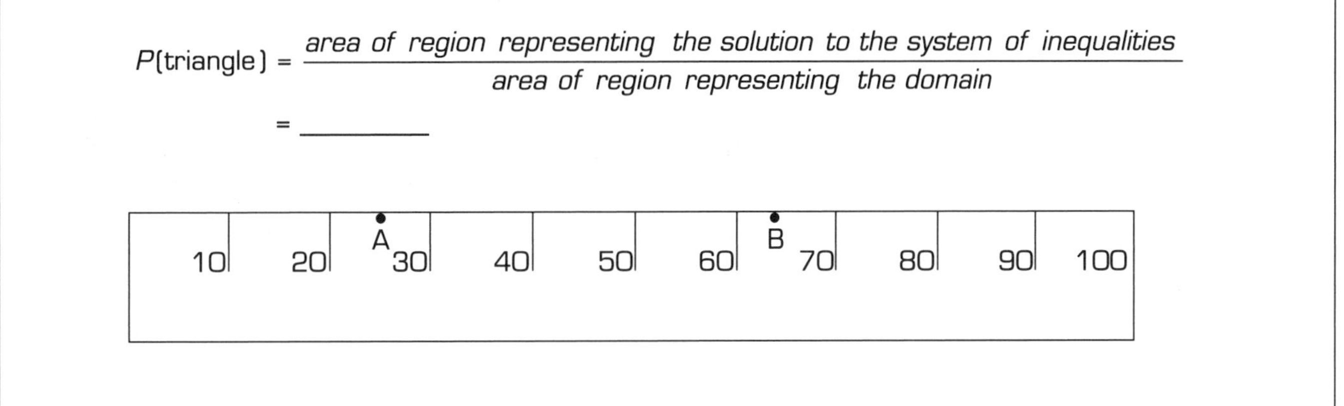

# CHAPTER 2
# AN INTRODUCTION TO DATA

"Pupils should be introduced to a wide variety of forms of graphical representations and use should be made of published information. Pupils should be encouraged to discuss information presented in diagrammatic form, especially in advertising." "Our aim should be to encourage a critical attitude toward statistics presented by the media."

Cockcroft Report 1982

*Teaching Matters:* **USA TODAY *is a particularly good source for graphical representations of real data.***

The *Curriculum and Evaluation Standards* emphasizes the ability to secure information from the many forms in which data are presented as one of the most important statistical skills. Newspapers, magazines, and even television display information in graphic and tabular forms. Graphic displays are favored by the media because they are the simplest and most powerful means of presentation. Not all graphics designers, however, are as knowledgeable as they should be about statistics or data analysis. Therefore, it is vital that students learn to read and evaluate that information. Students must be able to determine whether or not a presentation is consistent with the message contained in the data. Most graphical representations of data are not exploitations of the data, but students need to develop a critical attitude concerning the "graphical integrity" (Tufte 1983) of the presentation.

The material in this chapter, which can be used at any point in a ninth- or tenth-grade class, is designed to help students—

- extract information from data presented in graphs, charts, and tables;
- determine what these data say;
- communicate a message with a graphic display;
- determine which type of presentation is most appropriate for the message being presented in the data.

Formal teacher presentations are not necessary when using this material in the classroom. Instead, teachers need to guide and promote interaction as students share ideas and conclusions with one another. Communication is important; students need to learn to "express mathematical ideas orally and in writing," "read written presentations of mathematics with understanding," and "ask clarifying questions" about the content (NCTM 1989, p. 140). In order to promote communication, questions should be answered in complete sentences. Simple yes/no or numeric answers should not be accepted. Students will become aware that complete explanations are needed and will be more willing to give them as soon as it is apparent that questions can often be interpreted several ways, making it necessary to defend their conclusion with a reasonable and logical explanation.

*Teaching Matters:* ***When the students are seated in cooperative groups, answering questions or interjecting to aid and direct discussion should be done from a seated or kneeling position to avoid appearing as the authority.***

The nature of this material lends itself to the use of cooperative learning groups, which promote discussion and diverse interpretations and lead students to form opinions based on data. Teachers must "listen" to the discussion in each group to determine whether or not the students have discovered the distinction between different methods of solution. For example, is an estimate based on a visual image or a mathematical calculation? Students should be encouraged to clarify and extend their ideas. After the students have been working collectively on a question or series of questions, time should be allowed for the entire class to discuss their answers so they may come to realize that there are many correct interpretations of the same material.

Sharing with other teachers the difficulties students are having understanding the material extends the notion of cooperative learning to the staff. If possible, lessons can be coordinated with other departments to enable students to begin to transfer learning from one subject to another and to recognize that mathematics applies to many different fields. For example, activity 4 at the end of this chapter illustrates how mathematics is integral to an understanding of global education issues.

Of the five classroom-ready activities that follow, the first three provide opportunities for developing skills in understanding and interpreting data presented in tabular and graphic form. The remaining two activities provide opportunities for creating alternative representations for statistical information. Facility in working with percent is essential for each of the activities.

*Assessment Matters:* ***A series of interpretative questions about real-life situations that involve data, like those in the accompanying activities, can be the basis for an assessment of student understanding. On an hour-long test of this type, it is probably wise to include at least two different contexts. This lessens the chance that a student with a good general understanding of the mathematics will happen to have difficulty understanding the particular context on the test.***

***ACTIVITY 3***
***UNDERSTANDING DATA: POPULATION SHIFTS***

SHEET 1

The table below presents ranges for the predicted percent change in population of the United States from 1980 to 1990.

| State | Percent Change |
|---|---|
| Alabama | 5.0 to 9.9 |
| Alaska | 10.0 to 43.4 |
| Arizona | 10.0 to 43.4 |
| Arkansas | 5.0 to 9.9 |
| California | 10.0 to 43.4 |
| Colorado | 10.0 to 43.4 |
| Connecticut | 5.0 to 9.9 |
| Delaware | 10.0 to 43.4 |
| District of Columbia | 10.0 to 43.4 |
| Florida | 10.0 to 43.4 |
| Georgia | 10.0 to 43.4 |
| Hawaii | 10.0 to 43.4 |
| Idaho | 5.0 to 9.9 |
| Illinois | –5.3 to 4.9 |
| Indiana | –5.3 to 4.9 |
| Iowa | –5.3 to 4.9 |
| Kansas | 5.0 to 9.9 |
| Kentucky | –5.3 to 4.9 |
| Louisiana | 5.0 to 9.9 |
| Maine | 5.0 to 9.9 |
| Maryland | 10.0 to 43.4 |
| Massachusetts | –5.3 to 4.9 |
| Michigan | –5.3 to 4.9 |
| Minnesota | 5.0 to 9.9 |
| Mississippi | 5.0 to 9.9 |

| State | Percent Change |
|---|---|
| Missouri | 5.0 to 9.9 |
| Montana | –5.3 to 4.9 |
| Nebraska | –5.3 to 4.9 |
| Nevada | 10.0 to 43.4 |
| New Hampshire | 10.0 to 43.4 |
| New Jersey | 5.0 to 9.9 |
| New Mexico | 10.0 to 43.4 |
| New York | –5.3 to 4.9 |
| North Carolina | 10.0 to 43.4 |
| North Dakota | –5.3 to 4.9 |
| Ohio | –5.3 to 4.9 |
| Oklahoma | 5.0 to 9.9 |
| Oregon | 5.0 to 9.9 |
| Pennsylvania | –5.3 to 4.9 |
| Rhode Island | 5.0 to 9.9 |
| South Carolina | 10.0 to 43.4 |
| South Dakota | –5.3 to 4.9 |
| Tennessee | 5.0 to 9.9 |
| Texas | 10.0 to 43.4 |
| Utah | 10.0 to 43.4 |
| Vermont | 5.0 to 9.9 |
| Virginia | 10.0 to 43.4 |
| Washington | 10.0 to 43.4 |
| West Virginia | –5.3 to 4.9 |
| Wisconsin | –5.3 to 4.9 |
| Wyoming | 5.0 to 9.9 |

Source: U.S. Department of Commerce, Bureau of the Census, December 1989

Use the table to answer these questions:

1. The population of Alaska in 1980 was 303 000. What was the predicted population for 1990?
2. What does a negative percent mean?
3. Why are the percent values listed as a range and not as a single value?
4. What is the projected percent change in population in your home state?
5. How many states have a percent change from 5.0% to 9.9%?
6. What states are projected to have the largest percent gain in population?
7. Find out what the actual 1990 census figures were for your state. How close was the prediction?

**ACTIVITY 3** SHEET 2

**UNDERSTANDING DATA: POPULATION SHIFTS (CONTINUED)**

The following map is a graphical presentation of the same projected percent change in population for the 1990 census.

LEGEND
–5.3% to 4.9%
5.0% to 9.9%
10.0% to 43.4%

Use the map to answer the following questions:

8. What does the dotted area of this map represent?

9. How many states are included in each category listed in the legend?

10. What percent of the area of the continental United States would you *estimate* to be in the –5.3% to 4.9% range? How did you make your estimate?

11. What percent of the states of the entire United States is expected to have a population increase from 5.0% to 9.9%?

12. Does the projected population growth seem to be regional? Explain.

13. If the population of the state of Wisconsin was 4 705 642 persons, what is the maximum expected loss in population in the 1990 census? What is the maximum expected gain for the state? Would your state have a possible loss in population? Calculate the interval for the expected population growth in your state.

14. Why do you think the Census Bureau made these predictions knowing that they were actually going to take a census in 1990?

15. Do you prefer the table or the map as a way to present the information? Explain why you feel your choice is the better presentation.

16. Would the use of color on the graphic make a greater impression on the reader? Explain.

***ACTIVITY 3***

***UNDERSTANDING DATA: POPULATION SHIFTS (CONTINUED)***

SHEET 3

The following graph is often called a *pyramid graph*. It was also obtained from the U.S. Census Bureau and concerns the population distribution of the United States.

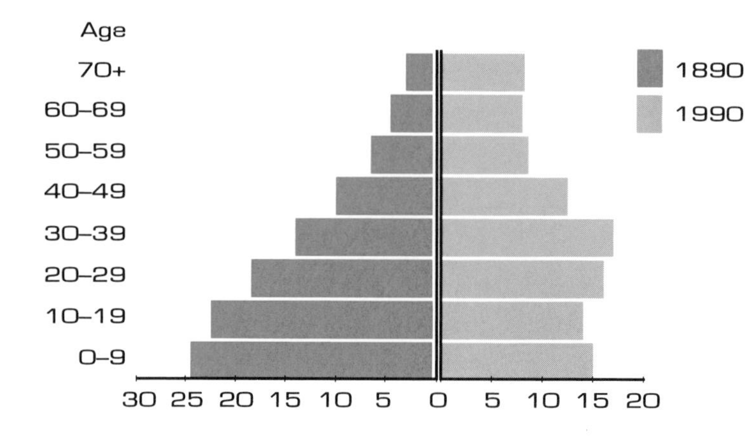


17. What percent of the population in 1890 was in the age range 30–39?

18. Compare the percent of the population in the age range 30–39 in 1890 to that in 1990. What conclusion can you make? What explanation might you give to support your conclusion?

19. What general observation would you make concerning the age distribution in 1990 compared to 1890? Justify your observation.

The two population pyramids below represent the age composition of the United States in 1985 and projected for the year 2030.

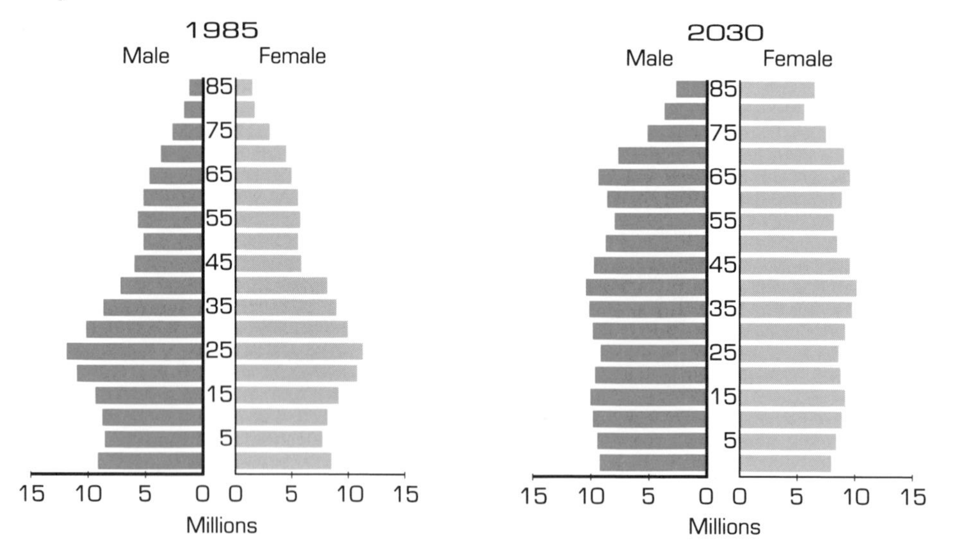


20. How many people in the 0–5 age bracket were there in the United States in 1985? How old will these people be in the year 2030?

21. How many people are predicted to be in that age bracket in the year 2030? What observations can you make and what might explain these observations?

22. Create a graph to represent the aging problem that is apparent from the two pyramid graphs pictured above and describe the problem it represents.

## *ACTIVITY 4*

SHEET 1

## *UNDERSTANDING DATA: POPULATION CHARACTERISTICS*

The map on sheet 2 is a graphic display of the probabilities of dying before the age of 5 in the years 1980–1985. After studying the map, answer the following questions:

1. Why is the size of the United States so small in comparison to India and China?
2. What is the probability of a child living in Brazil having a fifth birthday?
3. What is the meaning of the black regions on this map?
4. Give an explanation of why Bangladesh might have such a high probability of childhood death.
5. When people look at this information, they get the impression that infant death is a worldwide problem. Summarize why they get that impression.
6. Explain in your own words what a probability of 25 per 1000 means.

Another map of the same style is shown on sheet 3. This map presents data concerning the number of foreign-born persons living in the various countries of the world. After studying this information, answer these questions:

7. A legend in the lower right indicates the area representations. Explain how you should read that information.
8. What percent of the total population of France is foreign-born?
9. If the population of Poland in 1986 was estimated to be 37 546 000, approximately how many were foreign-born?
10. Which countries have populations of the fewest foreign-born? What are some possible reasons these countries have few foreign-born people?
11. What might cause a country to have a large population of foreign-born?
12. How would you determine what size to make a given country on this map?
13. Why would an individual country or the United Nations be interested in this information?
14. Why is this map made up of rectangular regions rather than the actual shape of the countries?

## ***UNDERSTANDING DATA: POPULATION CHARACTERISTICS (CONTINUED)***

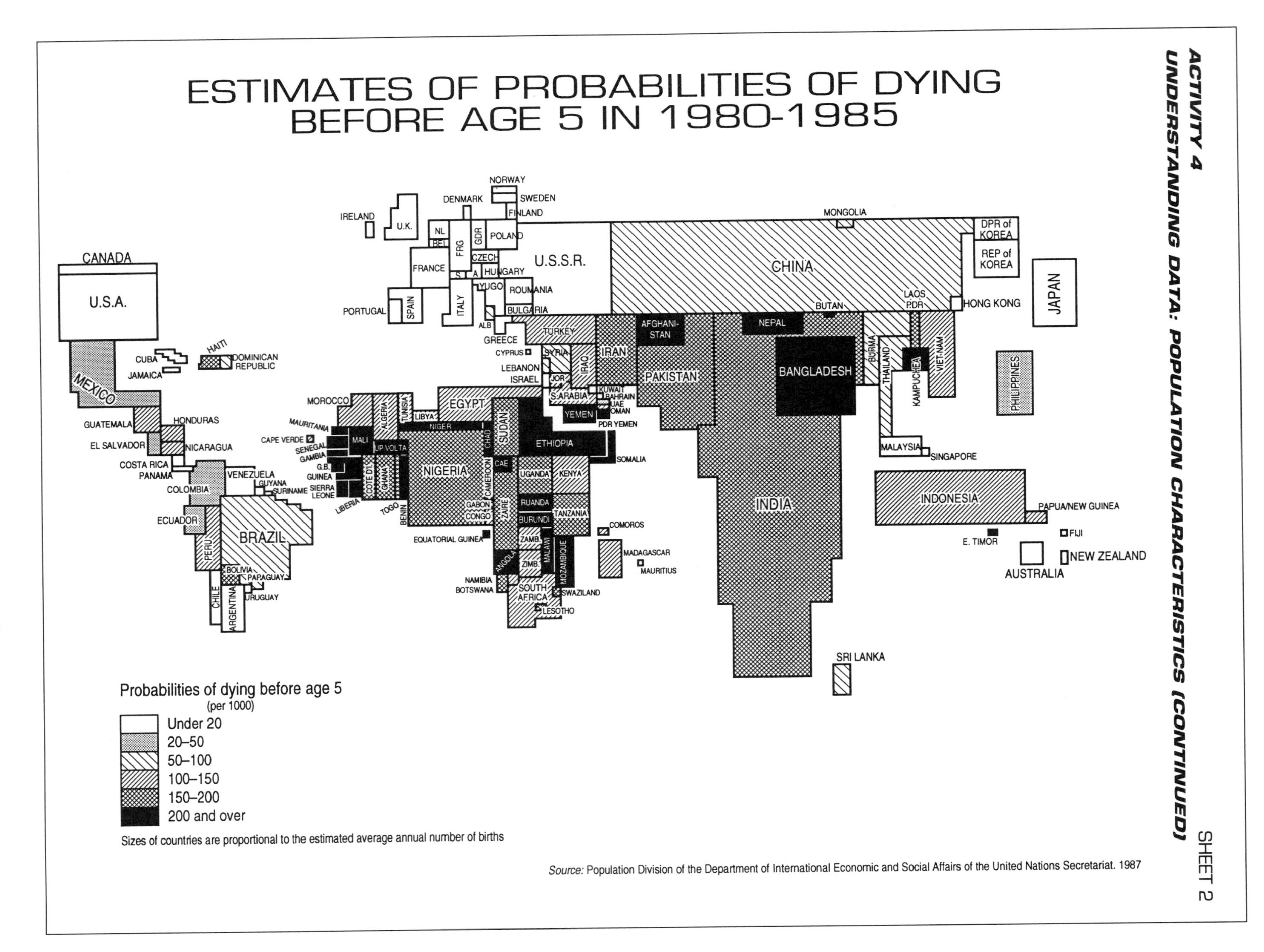

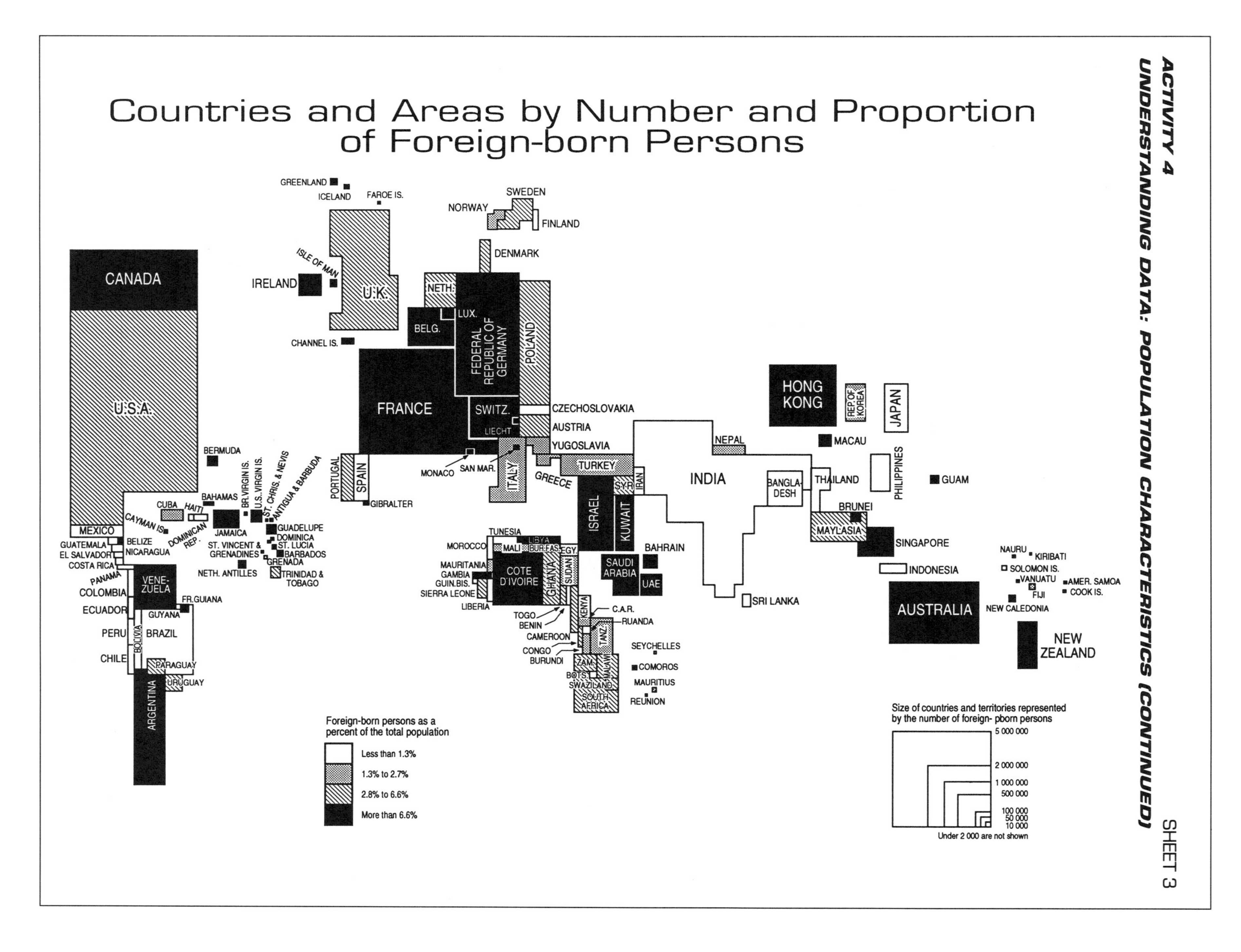
Countries and Areas by Number and Proportion of Foreign-born Persons
GREENLAND
ICELAND
FAROE IS.
NORWAY
SWEDEN
FINLAND
DENMARK
CANADA
U.S.A.
MEXICO
IRELAND
ISLE OF MAN
U.K.
CHANNEL IS.
NETH.
BELG.
LUX.
FEDERAL REPUBLIC OF GERMANY
POLAND
FRANCE
SWITZ.
LIECHT
CZECHOSLOVAKIA
AUSTRIA
YUGOSLAVIA
MONACO
SAN MAR.
ITALY
GREECE
TURKEY
PORTUGAL
SPAIN
GIBRALTER
BERMUDA
BAHAMAS
CUBA
HAITI
CAYMAN IS.
DOMINICAN REP.
JAMAICA
BR.VIRGIN IS.
U.S..VIRGIN IS.
ST. CHRIS. & NEVIS
ANTIGUA & BARBUDA
GUADELUPE
DOMINICA
ST. LUCIA
BARBADOS
GRENADA
ST. VINCENT & GRENADINES
NETH. ANTILLES
TRINIDAD & TOBAGO
GUATEMALA
BELIZE
EL SALVADOR
NICARAGUA
COSTA RICA
PANAMA
COLOMBIA
VENE-ZUELA
FR.GUIANA
GUYANA
ECUADOR
PERU
BOLIVIA
BRAZIL
CHILE
PARAGUAY
URUGUAY
ARGENTINA
TUNESIA
LIBYA
MOROCCO
MALI
BUR.FAS.
EGY
MAURITANIA
GAMBIA
GUIN.BIS.
SIERRA LEONE
LIBERIA
COTE D'IVOIRE
GHANA
SUDAN
TOGO
BENIN
CAMEROON
CONGO
BURUNDI
KENYA
C.A.R.
RUANDA
TANZ.
ZAM.
MALAWI
BOTS.
SWAZILAND
SOUTH AFRICA
SEYCHELLES
COMOROS
MAURITIUS
REUNION
ISRAEL
SYR
IRAN
KUWAIT
SAUDI ARABIA
BAHRAIN
UAE
NEPAL
INDIA
BANGLA-DESH
SRI LANKA
HONG KONG
REP.OF KOREA
JAPAN
MACAU
PHILIPPINES
THAILAND
BRUNEI
MAYLASIA
SINGAPORE
INDONESIA
GUAM
AUSTRALIA
NAURU
KIRIBATI
SOLOMON IS.
VANUATU
FIJI
AMER. SAMOA
COOK IS.
NEW CALEDONIA
NEW ZEALAND
Foreign-born persons as a percent of the total population
Less than 1.3%
1.3% to 2.7%
2.8% to 6.6%
More than 6.6%
Size of countries and territories represented by the number of foreign- pborn persons
5 000 000
2 000 000
1 000 000
500 000
100 000
50 000
10 000
Under 2 000 are not shown

## ***ACTIVITY 5***

SHEET 1

## ***UNDERSTANDING DATA: MEASURES OF SUCCESS***

The table gives the percent of people 25 years and older in the United States and the number of years of school they have completed.

1. In 1987, what percent of the people in the U.S. had at least a high school education?
2. In 1970, what percent of the people 25 and older had not completed a high school education?
3. If a random sample of 3000 people were selected in 1987, approximately how many would you expect to find who had not completed a high school education?

| | Percent of Population Completing School for | | | | | | |
|---|---|---|---|---|---|---|---|
| | 0–4 years | 5–7 years | 8 years | 9–11 years | 12 years | 13–15 years | 16 or more years |
| 1970 | 5.5 | 10.0 | 12.8 | 19.4 | 31.1 | 10.6 | 10.7 |
| 1980 | 3.6 | 6.7 | 8.0 | 15.3 | 34.6 | 15.7 | 16.2 |
| 1987 | 2.4 | 4.5 | 5.8 | 11.7 | 38.7 | 17.1 | 19.9 |

Source: *Statistical Abstract of the United States,* 1989

4. In 1970, what percent of the people who started college had finished?
5. In 1987, what percent of the people who started college had finished?
6. What trends can you observe from the data?

This table gives the *Professional Golfer's* top 25 male money winners for 1990.

| Golfer | Winnings $ |
|---|---|
| Paul Azinger | 944 731 |
| Ian Baker-Finch | 611 492 |
| Chip Beck | 571 816 |
| Mark Calcavecchia | 834 281 |
| Fred Couples | 757 999 |
| Steve Elkington | 548 564 |
| Jim Gallagher, Jr. | 476 706 |
| Wayne Grady | 527 185 |
| Hale Irwin | 838 249 |
| Peter Jacobsen | 547 279 |
| Tom Kite | 658 202 |
| Wayne Levi | 1 024 647 |
| Davis Love III | 537 172 |

| Golfer | Winnings $ |
|---|---|
| Bill Mayfair | 693 658 |
| Larry Mize | 668 198 |
| Gil Morgan | 702 629 |
| Jodie Mudd | 911 746 |
| Mark O'Meara | 707 175 |
| Greg Norman | 1 165 477 |
| Nick Price | 520 770 |
| Loren Roberts | 478 522 |
| Tim Sampson | 809 772 |
| Payne Stewart | 976 281 |
| Bob Tway | 495 862 |
| Lanny Wadkins | 673 433 |

Source: *USA TODAY*

Make a histogram of the men's earnings. Use a range of 450 000 to 1 200 000 with a scale of 50 000. Use your histogram to answer the following questions:

7. Estimate the percent of the top men golfers who made less than $850 000.
8. Estimate the mean winnings of the top PGA men.
9. Estimate the percent of the men golfers who earned from $550 000 to $950 000.
10. Estimate the salary that marked the top 50% of the winners. Compare this to the mean.
11. Describe a typical interval for the earnings of the top golfers.
12. Did any of the golfers earn an unusually high amount of money? What does this mean?

**ACTIVITY 6**

SHEET 1

**REPRESENTING DATA: DRINKING AND DRIVING**

Below is a graph that appeared on 4 September 1990 in *USA TODAY*. Write a short essay to be included with the graph explaining the data presented.

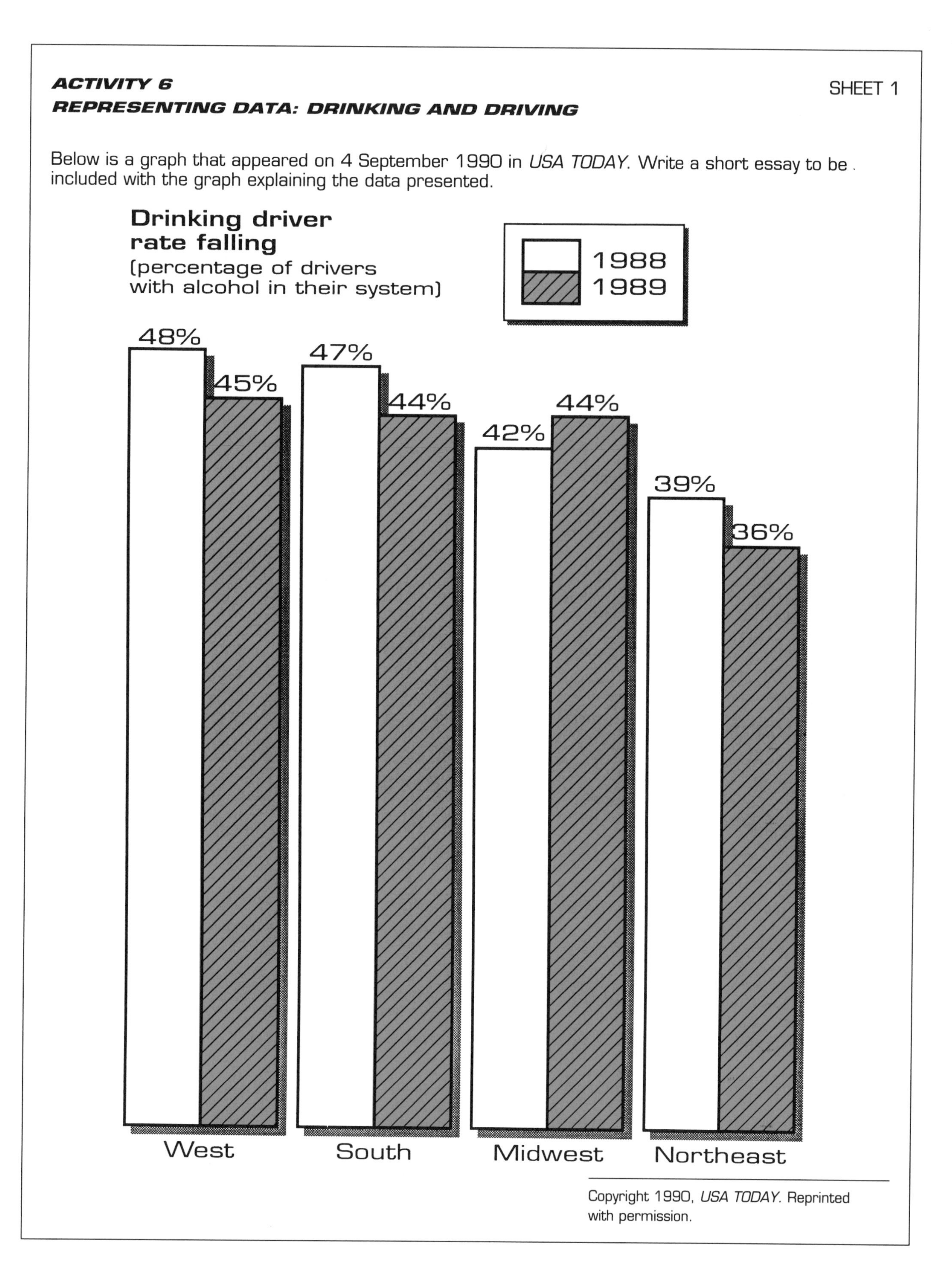

**ACTIVITY 7** SHEET 1

**REPRESENTING DATA: GRADES, ACTs, AND SATs**

The following article appeared in *USA Today*. Read the article and then create a graph or a chart that you feel would display the information conveyed as well as or better than just using words.

# A's in tough courses beat SATs, ACTs

How to get into college of your choice, 4D; Wednesday: Financial aid

By Pat Ordovensky
USA TODAY

High school grades, and the courses in which they're earned, are the most important factors in deciding who gets into college, say USA admissions directors.

Scores on college admissions exams—the SAT and ACT—rank third in importance among items on a student's application. Choosier colleges say a student's essay has greater weight than test scores.

So finds a USA TODAY survey of 472 admissions directors, selected randomly from four-year colleges.

The findings echo recent public statements by admissions officers that the SAT and ACT are losing value in predicting college success—that A's in tough high school courses are more relevant.

"Almost everyone keys on academic progress over 3 1/2 years" in deciding which applicants to admit, says Richard Steele, Duke University admissions director. "The combination of high grades and a quality high school program is the best predictor of success."

For the survey, USA TODAY talked to large private universities with well-known names and small state colleges that specialize in training teachers. Among them are 27 highly selective schools—from Harvard and Princeton to Stanford and Cal Tech—that accept an average of only 31 percent of their applicants. We found:

■ Almost 9 of 10 (88 percent) say grades are "very important" in admissions; 67 percent say rigor of high school courses; 52 say SAT/ACT scores; 45 percent say class rank.

■ 58 percent of all schools and 52 percent of the choosiest say grades are the "most important" of all factors. Second is tough high school courses, mentioned by 16 percent of all schools, 28 percent of the choosiest.

■ SAT and ACT scores are "most important" at 7 percent of the campuses but none of the selective schools.

■ 17 percent of all schools but 60 percent of the choosiest say an applicant's essay is a "very important" factor.

■ All schools responding have an average $1.5 million of their own money to help students who can't afford the tuition. At the selective schools, an average of $9 million is available.

■ 89 percent of the selective schools used a waiting list this year for qualified applicants who didn't make the final cut. Average number of names on the list: 396.

Brown University admissions director James Rogers, an author of the new book *50 College Admissions Directors Speak to Parents*, writes that experienced admissions officers "can review a transcript and predict the student's test scores."

And although women consistently score lower than men on SAT/ACT, 52 percent of this year's freshmen are women.

Other factors—recommendations, interviews—enter in, says Steele, because with talented applicants, "you don't eliminate large numbers if you stop at academic performance."

On less selective campuses, grades carry even more weight.

At 15 percent of the colleges responding, because of state law or school policy, SAT/ACT scores are deciding factors.

Three of every 10 schools surveyed (29 percent) say they're required to accept students meeting certain academic criteria. More than half (54 percent) say criteria include test scores.

The University of Arizona, for example, is required to accept any Arizona high school graduate who meets four requirements: a 2.5 (C-plus) grade average, 11 college prep courses, standing in top half of class, a 930 SAT score or 21 on the ACT.

Forty percent of Arizona's students are from out-of-state, says admissions director Jerry Lucido, but they need at least a 3.0 (B) high school average, a 1010 on the SAT or 23 ACT score.

At Michigan, North Carolina, Vermont and other public universities where more than twice as many non-residents apply as can be accepted, officials also say standards are much tougher for students who don't live in their states.

Among other factors considered important in the admissions process, the USA TODAY survey finds:

■ **Minority status:** Almost two-thirds (62 percent) of all schools and 96 percent of selective colleges say they're actively recruiting minority students who meet academic standards. That means minority status makes a difference between otherwise equal applications.

Of schools in the survey, this year's freshman class has an average 9 percent minorities—3 percent black.

■ **Recommendations:** 70 percent of all schools, 96 percent of the most selective, say recommendations from teachers and guidance counselors are important. More than half (52 percent) of the choosy schools say they're very important.

Other recommendations, from the community leaders, clergymen and such, are important to 52 percent of all schools, 75 percent of the choosiest.

But "choose your references carefully," says Steele. "More is not better.... Last year, an applicant gave us 23 unsolicited letters of recommendation. We were not impressed."

■ **Interviews:** Many highly selective schools use alumni across the USA to interview applicants in their area. At Duke, alumni reports weigh the same as teachers' recommendations.

Almost half (45 percent) of the surveyed schools and 64 percent of the choosiest say an interview is important.

■ **Essay:** It's important to 92 percent of the selective schools but only half (50 percent) of all schools.

It offers a clue to a student's "quality of thinking," says Steele. "Few kids can write their way in, but it can help."

Errors—factual and grammatical—hurt. At Boston University a few years ago, an applicant was rejected because his essay had too much Whiteout.

■ **Geography:** Most selective schools pride themselves on having students from all parts of the USA. A student from Montana applying to Harvard, for example, gets an edge over a New Englander.

At Duke, "we don't get many from the Dakotas," says Steele. "We snap to when one walks into the office."

More typical of all schools is Arizona where, Lucido says, "We're not shooting for geographic balance. We give no geographic preference."

■ **Alumni ties:** Admissions directors at selective schools say candidly that children of Alumni get special preference and children of generous alumni are even more special.

An alumni relationship is important to 44 percent of all schools, 80 percent of the most selective.

At Duke, 20 percent of all applicants are accepted. But for "alumni-connected" candidates the rate is over 40 percent.

# CHAPTER 3
# MAKING SENSE OF DATA

In a nation concerned about having the best, ratings have become very important. People often use ratings to decide which products to buy, movies or restaurants to go to, or corporations in which to invest.

Although some ratings are based on information that is objective, it is important to recognize that many aspects of ranking are very subjective, either arbitrary or dependent on someone's judgment. Whether or not you watch a TV show is objective; whether you liked the show or not is subjective. The number of touchdown passes a quarterback throws is objective; the way this number is factored into a rating is subjective. Usually the person doing the ranking and the ranking process have some credibility because of the credentials of the person or the organization. It is important, however, to remember that an expert in football, for example, might not be an expert in statistics. The way the data are put together to determine ranks is an important consideration in evaluating the results.

***Try This:* The Nielson ratings released each week list the top television shows. A rating of 21.1/32 means that 21.1 percent of all the 85.9 million households with television are watching that show. The 32 is called a share and means 32 percent of the televisions in use are tuned to the show. How are these ratings computed? How are these ratings used?**

The *Curriculum and Evaluation Standards* calls for a mathematics curriculum that focuses on reasoning, problem solving, communication, and connections. Analyzing data for purposes of ranking can illustrate problem solving in a real situation where reasoning and communication are vital. The process often links uses of algebra, geometry, and number sense and can be applied to problems from many different domains. Students can explore a real-world situation and "mathematicize" the problem to obtain a solution. There is frequently no single right answer, but there is a need for logical arguments that justify a solution. In this chapter we examine data to rank the "best" medium-priced car. We also examine the possibilities this set of data offers for integrating ideas from statistics with topics from more traditional mathematics. Related classroom-ready activity sheets appear at the end of the chapter.

The first step in making a decision based on a rank is to ask how that rank was computed. Were all the items ranked in the same way? What happens to "outliers"? Can an outlier change the ranks? Is one method of ranking better than another? What are some advantages and disadvantages of the ranking system?

***Teaching Matters:* Emphasize the critical issue of sampling in all statistical work. One complication in rating quarterbacks' prospects in tryout camp is the small sample of their performance that is actually observed. Even excellent performers have bad days, and relatively unskilled performers can look unduly talented for brief periods.**

In the December 1989 issue of *Car and Driver* magazine, staff members ranked eight cars, all priced under $10 000, to determine the "best" car. After a test drive, eight staff members subjectively rated the cars in each of eleven categories on a 1 to 5 scale with 5 being the best. These ratings were summed in each category for each car, thus making 40 the maximum possible rating for any category. The categories and the ratings are given in table 3.1.

How can these numbers be combined to choose the top-rated car? Several options might be considered.

- Find the total number of points for each car by summing the points from each category. The car with the highest number of points is the "best."
- To avoid the effect of outliers, eliminate the highest score and the lowest score for each car and then find the total number of points.
- In each category, assign a rank of 1 to the highest ranking car, 2 to the second highest, and so on. For example, under engine, the Honda

Table 3.1
Data Ratings

| Car | Engine | Trans-mission | Brakes | Handling | Ergo-nomics | Comfort | Ride | Utility | Styling | Value | Fun to drive |
|---|---|---|---|---|---|---|---|---|---|---|---|
| Ford Festiva LX | 22 | 28 | 27 | 22 | 29 | 27 | 22 | 27 | 25 | 35 | 23 |
| Honda Civic DX | 38 | 38 | 30 | 37 | 33 | 26 | 28 | 33 | 37 | 32 | 35 |
| Mazda 323SE | 28 | 26 | 22 | 26 | 30 | 28 | 30 | 29 | 23 | 26 | 24 |
| Mercury Tracer | 27 | 29 | 25 | 25 | 32 | 36 | 35 | 33 | 28 | 33 | 25 |
| Mitsubishi Mirage | 27 | 33 | 25 | 20 | 30 | 27 | 29 | 31 | 29 | 23 | 23 |
| Subaru Justy RS 4WD | 13 | 21 | 21 | 17 | 23 | 18 | 16 | 26 | 18 | 21 | 11 |
| Volkswagen Fox GL | 26 | 33 | 29 | 34 | 31 | 31 | 31 | 31 | 21 | 28 | 29 |
| Volkswagen Golf | 35 | 29 | 32 | 38 | 37 | 34 | 32 | 37 | 26 | 35 | 36 |

(Ergonomics is the accessibility and style of the seating and operational controls.)
Source: *Car and Driver*, December 1988

*Try This:* ***The Olympic diving competition is judged by seven judges, who rate each diver on a scale of 0 to 10. Find out how these ratings are combined to produce an Olympic medalist. Does it make a difference to the ratings if one country deliberately rates a performer high? In other words, will an "outlier" rating make a difference in the final ranking of the divers?***

Civic would have a 1 followed by the Volkswagen with a 2. Sum the rankings for each car. The best car is the car with the lowest number of points.

- The car with the best overall rating is the car with the most first places in the categories.

Table 3.2 gives the results of each method. The Volkswagen Golf seems to be the outstanding choice for the "best" car, first in each ranking system. In two categories, however, the Honda Civic was only 4 or 5 points behind, not a very definitive gap. Using the third method, the Mercury Tracer also looks as if it could be a contender. The cautious car buyer, however, might also have other concerns when looking at these ratings. What else might influence the selection of "best" car?

Table 3.2
Ranking Strategies

| Car | Total points | Points minus high, low | Ranked total | Number of firsts |
|---|---|---|---|---|
| Ford Festiva LX | 287 | 230 | 61 | 1 |
| Honda Civic DX | 367 | 303 | 30 | 3 |
| Mazda 323SE | 292 | 240 | 58 | 0 |
| Mercury Tracer | 328 | 268 | 35 | 2 |
| Mitsubishi Mirage | 297 | 244 | 53 | 0 |
| Subaru Justy RS 4WD | 205 | 168 | 88 | 0 |
| Volkswagen Fox GL | 324 | 270 | 43 | 0 |
| Volkswagen Golf | 371 | 308 | 20 | 6 |

***VARIABILITY***

One factor of interest in the rating process is the amount of variation in the ratings for each category. The same total could arise from adding two very high and two very low numbers or from adding four numbers fairly close together. Would you give a higher rank to the car with the most consistent ratings or to one whose ratings fluctuated wildly yet ended up with the same total?

Figure 3.1 shows box plots of the ratings for each car. (For information on making box plots, see *Exploring Data* [Landwehr and Watkins 1986] and "Exploring Data with Box Plots" [Bryan 1988].) Computer software such as Data Insights or Statistics Workshop, both from Sunburst Communications, will generate these graphs for students to analyze. From the plots, which car seems to have the most consistent ratings? Although the ratings for the Fox have the smallest range and interquartile range, they are not as high as the ratings for the Golf and the Civic. The Fox rating for style is an outlier; the Justy has inconsistent and low ratings. The Golf comes through again, although not by much over the Civic! The interquartile range for the Golf is slightly smaller and higher than for the Civic, although both have all ratings above 25.

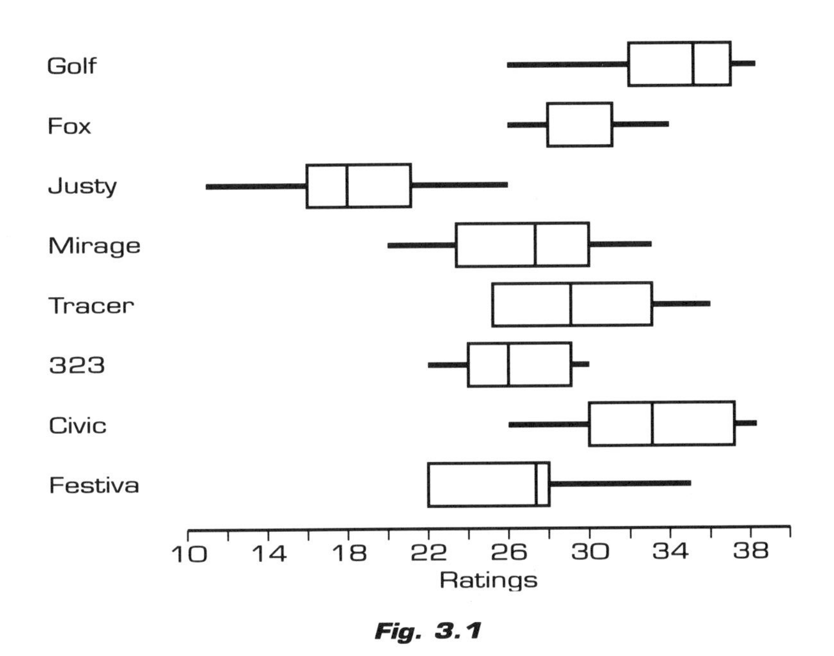


**Fig. 3.1**

*Teaching Matters:* ***All students should be able to create a rating scale and interpret the information using their scales and graphs. With appropriate software, students can analyze distributions and plots to compare the rankings in two categories.***

### RATING FROM A DIFFERENT PERSPECTIVE

Rating a car means different things to different people. Although many people are interested in value and utility, others are more concerned with engine and transmission or with styling and comfort. One possible way to accommodate different preferences is to weight these categories in finding the total result. Another method, which shows the relationship visually, is to make a scatterplot of the two variables concerned. The line $y = x$ will represent every car with equal ratings in the two categories. The car highest in both categories will be the car farthest from the origin and closest to the line $y = x$.

*Try This:* ***Have students find the average "error" by measuring the vertical distance from each point to the line in figure 3.2 and calculating an average distance.***

Figure 3.2 displays the ratings for comfort (C) and style (S). How many cars have a higher rating for comfort than they do for style? In algebraic terms, for how many points is C > S? (This is an excellent example of graphing inequalities and provides a meaningful basis on which students can build.) There are no cars in the upper right-hand corner of the graph with high ratings in both comfort and style. It seems cars with style are not comfortable!

*Try This:* ***Have precalculus students use the normal form of an equation of the line to calculate the directed distance from the points to the line in figure 3.2 and use these distances to analyze how well the line summarizes the data.***

Figure 3.3 is a scatterplot of utility and value. Which car is the highest in both of these ratings? It seems as if the answer could be either the Golf or the Tracer. One way to explore this further is to use algebra and geometry. Visualizing a graph can lead to some conjectures; coordinate methods can be used to justify or disprove those conjectures. Because the optimum rating in each category is 40, the situation requires finding the point that is closest to (40, 40). The distance between two points

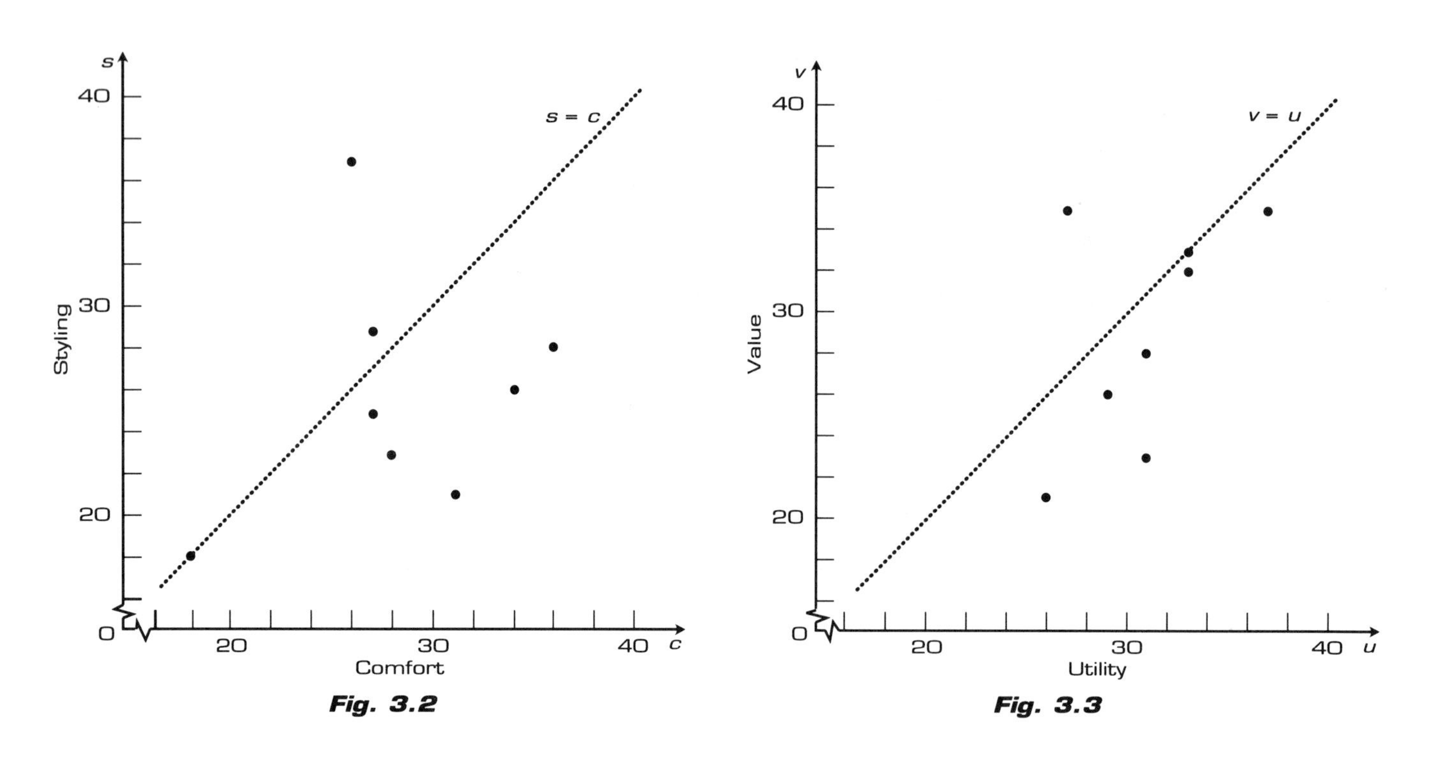


**Fig. 3.2**

**Fig. 3.3**

can be found by using the distance formula, $d = \sqrt{(x_1 - x_2)^2 + (y_1 - y_2)^2}$, where $(x_1, y_1)$ and $(x_2, y_2)$ are the coordinates of any two points in the plane. For example, the Golf's ratings for utility and value are 37 and 35 respectively. The distance from (37, 35) to (40, 40) is as follows:

***Try This:* Have students build a three-dimensional model and represent the ordered triples for three categories for each car. Which point would represent the optimum ratings for those three categories? Which car is closest to this point? How can you tell?**

$$d = \sqrt{(37 - 40)^2 + (35 - 40)^2}$$
$$= \sqrt{9 + 25} = \sqrt{34}$$
$$\approx 5.83$$

Using the same process, we find that the distance between the point representing the Tracer's rating and the maximum point is $\sqrt{18} \approx 4.24$. Thus, the Tracer rates higher in both categories than the Golf.

Now suppose three values, engine, transmission, and brakes, are important. What kind of geometric model could be used to find the car that ranked highest in all three? Construct a three-dimensional graph such as figure 3.4 and plot an ordered triple for each car for the three variables involved. Consider the point *A* (40, 40, 40) the optimum for the three variables, and analyze each car according to the distance of the point determined by their ratings from *A*. The "best" car in terms of all three qualities is the car closest to vertex *A*. It looks as if the Golf is the best car for engine, transmission, and brakes. Let's investigate the conclusion using coordinate methods. The Festiva was rated 22, 28, 27 for engine, transmission, and brakes respectively. This information is represented by the point *F* (22, 28, 27) in the graph.

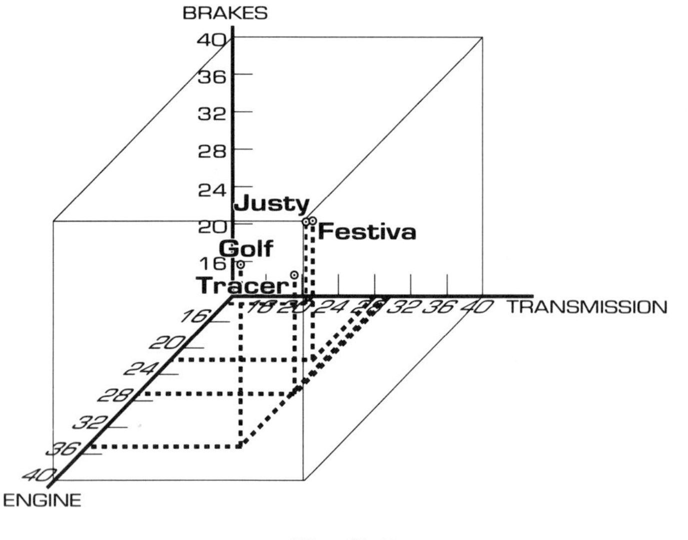


**Fig. 3.4**

The distance formula in three dimensions is

$$d = \sqrt{(x_1 - x_2)^2 + (y_1 - y_2)^2 + (z_1 - z_2)^2},$$

where $(x_1, y_1, z_1)$ and $(x_2, y_2, z_2)$ are the coordinates of any two points. Thus, the distance from $A$ to $F$ is

$$d = \sqrt{(40 - 22)^2 + (40 - 28)^2 + (40 - 27)^2} = \sqrt{324 + 144 + 169}$$
$$= \sqrt{637} \approx 25.23.$$

Table 3.3 shows the distance of each of the cars from the optimum rating. A car buyer who is looking for top quality in engine performance, transmission, and brakes and who is relying on the ratings by the *Car and Driver* staff should select the Honda Civic. Its distance from the maximum rating is only 10.4 ($\sqrt{108}$) units. Note that the distances were distorted by the graph, and thus using only the picture would have led to the wrong conclusion.

Table 3.3
Using a Three-dimensional Model

| Car | Distance from (40, 40, 40) | | | | | | |
|---|---|---|---|---|---|---|---|
| Ford Festiva LX | $\sqrt{(40-22)^2+(40-28)^2+(40-27)^2}$ | = | $\sqrt{324+144+169}$ | = | $\sqrt{637}$ | ≈ | 25.2 |
| Honda Civic DX | $\sqrt{(40-38)^2+(40-38)^2+(40-30)^2}$ | = | $\sqrt{4+4+100}$ | = | $\sqrt{108}$ | ≈ | 10.4 |
| Mazda 323SE | $\sqrt{(40-28)^2+(40-26)^2+(40-22)^2}$ | = | $\sqrt{144+196+324}$ | = | $\sqrt{664}$ | ≈ | 25.8 |
| Mercury Tracer | $\sqrt{(40-27)^2+(40-29)^2+(40-25)^2}$ | = | $\sqrt{169+121+225}$ | = | $\sqrt{225}$ | = | 15 |
| Mitsubishi Mirage | $\sqrt{(40-27)^2+(40-33)^2+(40-25)^2}$ | = | $\sqrt{169+289+225}$ | = | $\sqrt{683}$ | ≈ | 26.1 |
| Subaru Justy RS 4WD | $\sqrt{(40-13)^2+(40-21)^2+(40-21)^2}$ | = | $\sqrt{729+361+361}$ | = | $\sqrt{1451}$ | ≈ | 21.2 |
| Volkswagen Fox GL | $\sqrt{(40-26)^2+(40-33)^2+(40-29)^2}$ | = | $\sqrt{196+289+121}$ | = | $\sqrt{606}$ | ≈ | 24.6 |
| Volkswagen Golf | $\sqrt{(40-35)^2+(40-29)^2+(40-32)^2}$ | = | $\sqrt{25+121+64}$ | = | $\sqrt{210}$ | ≈ | 14.5 |

If someone were interested in four qualities, a graphic representation is no longer practical, but the coordinate model can easily be extended to include four or as many dimensions as desired.

***THE FACTS***

In addition to making subjective judgments about the qualities of a car, buyers usually request certain factual data before they will decide which car to buy. Table 3.4 displays this information for the eight cars tested. Acceleration is the number of seconds it takes a car to go from 0 to 60 mph, braking is the number of feet it takes a car traveling at 70 mph to stop, EPA is the fuel economy in miles per gallon, and the cost is the actual cost of the car tested, not necessarily the base price.

Would it be appropriate to find a total by adding the numbers in each category, as was done earlier? In table 3.1 large numbers meant high ratings in each category. Now, however, small numbers are more desirable in acceleration, braking, and cost; large numbers are better in miles per gallon, making a total sum meaningless. One way to compensate for this is to apply the second method used earlier; give every car a rank in each category and then sum the ranks for each car. These results are given in table 3.5. How do they compare to the results obtained when the same process was used in table 3.1?

Teaching Matters: ***Repetitive calculations as in table 3.3 provide opportunities for students to construct and implement simple computer algorithms.***

Assessment Matters: ***Computer software (such as spreadsheets) and graphing calculators would be useful tools for activities described in this section. Design tests that require the use of these tools.***

Table 3.4
Factual Information

| Car | Acceleration sec. (0–60 mph) | Braking ft. (70–0 mph) | EPA city | EPA highway | 700-mi. trip | Cost |
|---|---|---|---|---|---|---|
| Ford Festiva LX | 12.5 | 193 | 39 | 43 | 33 | 8382 |
| Honda Civic DX | 9.6 | 206 | 33 | 37 | 30 | 9975 |
| Mazda 323 SE | 10.6 | 199 | 28 | 35 | 31 | 9954 |
| Mercury Tracer | 10.8 | 225 | 28 | 35 | 29 | 9631 |
| Mitsubishi Mirage | 12.3 | 209 | 32 | 37 | 33 | 9929 |
| Subaru Justy RS 4WD | 12.4 | 218 | 30 | 35 | 29 | 9144 |
| Volkswagen Fox GL | 12.3 | 205 | 24 | 29 | 27 | 9695 |
| Volkswagen Golf | 9.6 | 200 | 25 | 33 | 30 | 9870 |

Table 3.5
Comparison of Ratings

| Car | Ratings from C/D staffers | Rank | Ratings from facts | Rank |
|---|---|---|---|---|
| Ford Festiva LX | 61 | 7 | 13 | 1 |
| Honda Civic DX | 30 | 2 | 22 | 2 |
| Mazda 323 SE | 58 | 6 | 24 | 4 |
| Mercury Tracer | 35 | 3 | 30 | 6 |
| Mitsubishi Mirage | 55 | 5 | 23 | 3 |
| Subaru Justy RS 4WD | 88 | 8 | 30 | 6 |
| Volkswagen Fox GL | 43 | 4 | 37 | 8 |
| Volkswagen Golf | 20 | 1 | 27 | 5 |

The graph in figure 3.5 represents a comparison of the two rankings. C/D staff rankings are denoted by *, and the rankings based on facts are denoted by •. The results are surprising. The Ford Festiva changed from a rank of 7 to 1 (note the length of the line connecting the two ratings). The only car with the same rank for both sets of data was the Honda Civic. The Volkswagen Golf, number 1 in the first analysis, was repositioned to fifth on the basis of some hard facts. The buyer must still determine if all the facts are equally important.

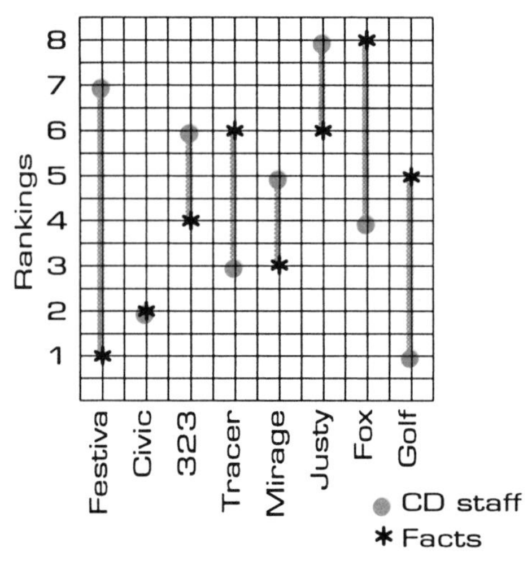


***Fig. 3.5***

Which car would you rate as best using the numbers provided by the staff of *Car and Driver?* For the first data set, the top point earner was the Volkswagen Golf, but only by a 4-point margin over the Honda Civic. When the cars are compared to each other, category by category the best car was the Golf with a wider margin over the Civic. The Golf and the Civic are about equal in terms of consistently high ratings. For value and utility, the Golf is out in front again, although the Tracer and the Civic are close behind. If engine, transmission, and brakes are important, the top car is the Honda Civic. To the driver who values comfort, style, and ride, however, the winner will be the Mercury Tracer! On the basis of objective data, the Ford Festiva is number 1, with the Honda Civic in second place. Which is more important, ratings in terms of style and handling or miles per gallon? Which car would you buy?

When data are used to present a case or as the basis for a decision, it is important for students to understand what those data mean. Who gathered the data and how? What process was used to organize the data? What was important in the organization and display of the data? What assumptions underlie the work? Are those assumptions consistent with your needs and conditions? Sometimes the data obviously support the hypothesis or argument. More often, however, there are many ways to look at the data. Wise users of statistics recognize this as they look at the ratings for the "best" community in which to live or the ten best plays of the year or the best medium-priced car to buy.

*Teaching Matters:* ***Students in second-year algebra and precalculus can determine whether two variables are related by using a calculator to find the correlation coefficient. For example, plot the number of miles per gallon in the city and on the highway for a group of cars and check the correlation coefficient against the graph.***

*Try This:* ***An important rating for many people is the frequency and availability of repair and service for cars. Have students find out from the librarian where this information can be obtained and prepare a report for the class.***

SHEET 1

## *ACTIVITY 8*
## *THE RATING GAME: WHICH CAR IS THE "BEST"?*

The table below gives combined ratings of eight staff members of *Car and Driver* magazine in a test of compact automobiles.

Data Ratings

| Car | Engine | Trans-mission | Brakes | Handling | Ergo-nomics | Comfort | Ride | Utility | Styling | Value | Fun to drive |
|---|---|---|---|---|---|---|---|---|---|---|---|
| Ford Festiva LX | 22 | 28 | 27 | 22 | 29 | 27 | 22 | 27 | 25 | 35 | 23 |
| Honda Civic DX | 38 | 38 | 30 | 37 | 33 | 26 | 28 | 33 | 37 | 32 | 35 |
| Mazda 323SE | 28 | 26 | 22 | 26 | 30 | 28 | 30 | 29 | 23 | 26 | 24 |
| Mercury Tracer | 27 | 29 | 25 | 25 | 32 | 36 | 35 | 33 | 28 | 33 | 25 |
| Mitsubishi Mirage | 27 | 33 | 25 | 20 | 30 | 27 | 29 | 31 | 29 | 23 | 23 |
| Subaru Justy RS 4WD | 13 | 21 | 21 | 17 | 23 | 18 | 16 | 26 | 18 | 21 | 11 |
| Volkswagen Fox GL | 26 | 33 | 29 | 34 | 31 | 31 | 31 | 31 | 21 | 28 | 29 |
| Volkswagen Golf | 35 | 29 | 32 | 38 | 37 | 34 | 32 | 37 | 26 | 35 | 36 |

(Ergonomics is the accessibility and style of the seating and operational controls.)
Source: *Car and Driver*, December 1988

1. Use the grid provided to make a scatterplot of the ratings for transmission versus the ratings for handling. Put the ratings for transmission on the horizontal axis. Use the plot to answer these questions:
   a. In general, do cars rate higher in transmission or in handling? ______
      Explain how the graph helps you arrive at an answer. ______
   b. Do any of the points seem to cluster or group together? ______
      Describe the cars in each group. ______
   c. Which car rates highest in both of these categories? ______
      Do you need the distance formula to justify your choice? Explain. ______

2. Make a scatterplot on the grid provided of the ratings for brakes versus handling. Put the ratings for brakes on the horizontal axis. Use the plot to answer these questions:
   a. Is there a relationship between brakes and handling? ______
      If so, describe it. ______
   b. Which car rates the highest in both of the categories? ______
      Use the distance formula to verify your choice. Explain. ______

***ACTIVITY 8***
***THE RATING GAME: WHICH CAR IS THE "BEST"? (CONTINUED)***

SHEET 2

3. Use the distance formula to determine which car has the highest rating using brakes, handling, and transmission.

_______________

_______________

_______________

4. Let $E$ be the point representing the engine rating for the Civic (38) and the Golf (35), that is, the point $E(38, 35)$. Let $T$ be the point representing the transmission rating for the Civic and the Golf (38, 29). Plot $E$, $T$, and the points representing the paired ratings for each of the other categories on the grid provided. Use the plot to answer these questions:

   a. For which categories did the Civic rate higher than the Golf? _______________

   b. For which categories are the ratings farthest apart? _______________

   The closest? _______________

   c. Use the plot to determine which of these two cars you would select as best. Justify your answer.

   _______________

5. Using the information in this table as well as that from the table on sheet 1, decide which is the best car. Use at least two graphic displays and two numerical summaries in providing an argument for your claim.

Factual Information

| Car | Acceleration sec. (0–60 mph) | Braking ft. (70–0 mph) | EPA city | EPA highway | 700-mi. trip | Cost |
|---|---|---|---|---|---|---|
| Ford Festiva LX | 12.5 | 193 | 39 | 43 | 33 | 8382 |
| Honda Civic DX | 9.6 | 206 | 33 | 37 | 30 | 9975 |
| Mazda 323 SE | 10.6 | 199 | 28 | 35 | 31 | 9954 |
| Mercury Tracer | 10.8 | 225 | 28 | 35 | 29 | 9631 |
| Mitsubishi Mirage | 12.3 | 209 | 32 | 37 | 33 | 9929 |
| Subaru Justy RS 4WD | 12.4 | 218 | 30 | 35 | 29 | 9144 |
| Volkswagen Fox | 12.3 | 205 | 24 | 29 | 27 | 9695 |
| Volkswagen Golf | 9.6 | 200 | 25 | 33 | 30 | 9870 |

Source: *Car and Driver*, December 1988

6. If you were a Honda dealer, how would you use the data to advertise? _______________

_______________

_______________

7. If you were a Subaru dealer, how would you use the data to advertise? _______________

_______________

_______________

SHEET 1

## ***ACTIVITY 9***
## ***EXPLORING RATINGS***

1. Identify two areas where rating is used to select the best entry. How are these ratings made? In your opinion, are the methods "fair"?

2. Suppose there were six papers selected for the final round of a competition. Five judges each rated their top three papers as 1, 2, and 3, with 1 as the best paper. The results were as follows:

| | |
|---|---|
| Paper A | 2, 1, 3, 2, 2 |
| Paper B | 3 |
| Paper C | 1, 1, 1, 3 |
| Paper D | 2, 3, 2 |
| Paper E | 1, 3 |
| Paper F | (did not place) |

   If first place receives $5000 and second place receives $1000, which two papers do you think should receive the prizes and why? Be prepared to defend your decision to the rest of the class.

3. Set up your own rating scheme using categories you define and classmates as "experts." Possible rankings might be done using baseball players, actors, rock groups, or quarterbacks.

4. Collect the following information from your classmates: male or female; grade point average; and average number of hours per week spent on these activities: studying, working at an outside job, watching television, practicing an extra curricular activity (music, basketball, soccer, etc.), talking on the phone.

   Use scatterplots and the line $y = x$ to answer the following questions. Use different colors to record the information for the boys and for the girls.

   a. Who studies the most and watches the most television?

   b. Who studies the least and watches the most television?

   c. Who works the most and studies the most?

   d. Who practices the most and studies the least?

   e. Who talks on the phone the most and watches the least television?

   f. Who works the most and studies the least?

   g. Show how the distance formula can be used to verify your answers.

   h. Use the distance formula for three points to determine who has the highest grade point average, practices the most, and studies the most. (You will have to create an "optimum" number of hours of practice and study in order to compare the data.)

## CHAPTER 4
## EXPLORING LINEAR DATA

The study of functions is central to a curriculum that aligns with the *Curriculum and Evaluation Standards*. Students should be able to model real-world phenomena with a variety of functions and represent data symbolically, graphically, and in tables (NCTM 1989, p. 154). In earlier chapters general statistical concepts were investigated and some appropriate activities suggested. This chapter focuses on connections between statistics and functions strands by graphing points representing real-world data and modeling the observed linear relations by linear equations. These topics are part of the present prealgebra–algebra I–geometry–algebra II–precalculus curriculum, yet many students in calculus have not yet mastered even the concept of slope. The use of activities based on data in earlier courses may make linear equations more meaningful for students and promote a better grasp of the topic by them.

*Teaching Matters:* ***It is important that students come to understand that—***

- ***for some data sets, the assignment of a variable to an axis is purely arbitrary;***
- ***the spread of data about a line can be measured by the correlation coefficient, and the smaller the value of the correlation coefficient, the more important it is to reverse the regression to make an accurate prediction.***

Again the first step is to understand the data. For some relations there is clearly an independent, or operating, variable and a dependent, or response, variable—for example, time and distance. The choice when fitting lines does not always depend on the physical relation between the operating and response variables. It is often more important to know which of the two is to be predicted. If, for example, students investigate the relationship between the temperature and the number of times a cricket chirps in a given period of time, there is a linear relationship. Clearly, the temperature determines to a large extent the rapidity with which crickets chirp, not the other way around. If the students want to predict the temperature by counting chirps, however, the better prediction would come from using the number of chirps as the independent variable and the temperature as the dependent variable.

Table 4.1 displays data that relate the number of oil changes per year and the cost of engine repairs. The activity "Oil Changes and Engine Repairs" at the end of this chapter uses these data to introduce students to modeling with a linear function. To predict the cost of repairs from the number of oil changes, use the number of oil changes as the $x$ variable and engine-repair cost as the $y$ variable. The axes should be labeled, including the units (oil changes/year and dollars), and marked so that all the ordered pairs in the table can be plotted.

Table 4.1

| Oil changes per year | 3 | 5 | 2 | 3 | 1 | 4 | 6 | 4 | 3 | 2 | 0 | 10 | 7 |
|---|---|---|---|---|---|---|---|---|---|---|---|---|---|
| Cost of repairs | $300 | 300 | 500 | 400 | 700 | 400 | 100 | 250 | 450 | 650 | 600 | 0 | 150 |

Figure 4.1 displays the data from table 4.1 graphically. The students are asked to visualize a straight line as a representation of the data. Each student should draw a line that seems to "fit" the plotted points. A line that "fits" the points should have the same characteristics as the set of points; it should actually summarize the data. A line drawn this way is called an eyeball-fit line.

*Teaching Matters:* ***Initially students can make an eyeball line; later they can find a median-fit line or determine a regression line by using a graphing calculator, spreadsheet, or data analysis software.***

Other fitted lines, such as the median-fit line (Landwehr and Watkins 1986; Shulte and Swift 1986) and the least squares regression line (North Carolina School of Mathematics and Science 1988), can be constructed depending on the level of the students. Students can compare their lines, which should differ only slightly. All the lines should share one characteristic: they slope downward. This activity, so far, is appropri-

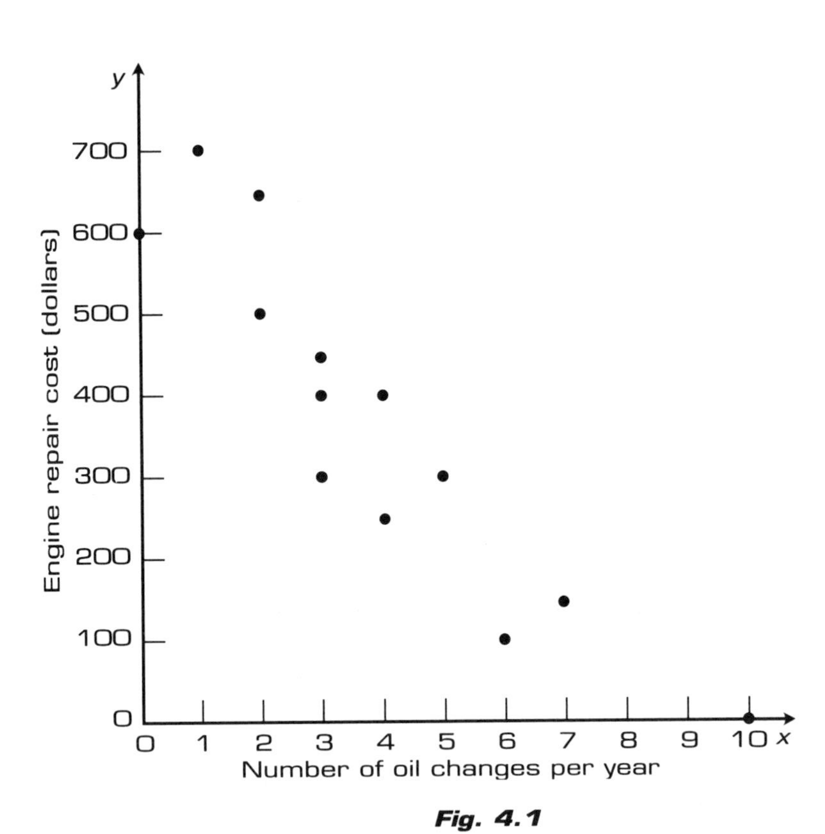


***Fig. 4.1***

*Teaching Matters:* ***Limitations of using the "best fit" line to predict future data should be discussed. For example, a best-fit line for the winning Olympic times in the 100-meter dash would predict a negative winning time within twenty years, but of course this is impossible. Nonlinear data are discussed in the next chapter.***

*Teaching Matters:* ***Emphasize that the intercepts of a fitted line do not make sense for all problems. For example, in Activity 11, when the height is 0, the weight cannot be a positive number! Be sure students consider the domain and range implied by the problem context.***

ate for a middle grades class as well as a high school class. The concept of slope can be introduced early in a student's school career if it is approached through an activity based on data.

What is the significance of the line's downward (or negative) slope? Students will say it means that the more oil changes made per year, the less money spent on engine repairs. Note that the correct wording should be, There is a *relation* between the number of oil changes and the money spent on repairs. Although the line represents the data, there is no indication that the number of oil changes causes the need for engine repairs. There are many other variables that affect the cost of engine repairs. More oil changes may just mean those cars have more careful drivers.

Students should examine the slope, which can be found by counting units. First, they must define what a unit represents for each variable. In figure 4.1, one horizontal unit represents one oil change. One vertical unit represents 100 dollars in engine repairs. Since students have drawn different lines, the slopes will vary. Note that whereas a fraction or ratio is easily understood as the slope or rate of change, a decimal representation is more useful to compare slopes. After listing all the slopes reported by the class (perhaps in a stem-and-leaf plot), the class should determine a "consensus" slope. The number should be simple to use. For example, –100 is much more useful than –98.73. Since utility is more highly valued than precision in this instance, a slope of –100 seems to be a reasonable answer.

Students should be able to interpret slope as rate of change. What is the change in the cost of repairs for each oil change? A rate of change or slope of –100/1 indicates that for each additional oil change per year, the cost of engine repairs will *tend* to *decrease* by 100 dollars. Associating measurement units with the slope is important to give students a concrete basis for understanding. Students should recognize that changing the units on an axis will affect the slope. If the vertical axis were in cents rather than dollars, the slope would be –10 000.

Next examine the intercepts. The *y*-intercept is about 750. This means that if there are zero oil changes, engine repairs will cost about $750. From the graph the *x*-intercept is about 7.5, which means that a car owner would expect to spend nothing on engine repairs if she changed the oil 7.5 times a year. Is this a sensible number of oil changes per year?

The slope and the *y*-intercept can be used to write the equation of the line:

Writing $y = mx + b$ for $m = -100$ and $b = 750$, we get $y = -100x + 750$.

If the *y*-intercept is not accessible because of the scale, the equation of the line can be found by using any two points (not necessarily data points) on the line. Students can use the equation to predict the cost of engine repairs expected for a specific number of oil changes. For example, if you change your oil four times a year, how much can you expect to pay in engine repairs? Let $x = 4$, then $y = -100(4) + 750$, or $y = \$350$.

You can expect to pay about $350 in engine repairs.

What is the difference between data points above the line and those below the line? As mentioned in chapter 1, the concept developed here is very helpful for graphing inequalities. Points above the line would indicate that the actual engine repairs exceeded the amount predicted by the number of oil changes. Points below the line represent situations where the engine repairs cost less than predicted. Because the line is only a summary of the relation, just as the mean or median is a summary for a single set of data, there is a degree of variation in using the line to predict. Additional discussion could explore possible reasons, in addition to the natural variation, for this deviation from the line. Excessive engine repairs could be due to bad driving habits; lower than expected repair costs could be due to a special type of oil.

Students can obtain data for further exercises from the newspaper or reference books or use data actually collected from other students. They need to remember that not all data are linear. Before a fitted line is drawn, the most important question to ask is, "Could you imagine a line that represents these points?" Using real data to teach linear equations makes the material relevant and concrete. The last four activities at the end of this chapter provide additional rich contexts for relating linear equations and data.

*Try This:* ***Have students write a computer program that generates data to roughly fit a line, say $y = 2x + 1$, by randomly generating y-values between $2x$ and $2x + 2$ for each integer between 1 and 10 inclusive. The result will help them understand the concept of a "best fit" line for real data.***

*Try This:* ***Have students use data analysis software to create a median-fit line for their own set of data. Analyze this line in terms of slope and intercept. Discuss how well the line "fits" the data.***

*Try This:* ***Have students measure their height and the height of their waist from the floor. Plot the data and analyze the linear relation.***

***ACTIVITY 10***

***OIL CHANGES AND ENGINE REPAIR***

SHEET 1

1. The table gives data relating the number of oil changes per year to the cost of car repairs. Plot the data on the grid provided, with the number of oil changes on the horizontal axis.

| Oil changes per year | 3 | 5 | 2 | 3 | 1 | 4 | 6 | 4 | 3 | 2 | 0 | 10 | 7 |
|---|---|---|---|---|---|---|---|---|---|---|---|---|---|
| Cost of repairs | $300 | 300 | 500 | 400 | 700 | 400 | 100 | 250 | 450 | 650 | 600 | 0 | 150 |

2. Are the data linear? If so, draw a best-fit line.

3. Find the slope of the line. Describe in words what the slope represents.

4. Find the *x*- and *y*-intercepts. Explain in terms of oil changes and engine repairs what each represents.

5. Write the equation of the line.

6. Use the equation to predict the cost of engine repairs if the car had four oil changes. How accurate do you think your prediction is? Explain your answer.

***ACTIVITY 11***
***BIKE WEIGHTS AND JUMP HEIGHTS***

SHEET 1

1. In BMX dirt-bike racing, jumping high or "getting air" depends on many factors: the rider's skill, the angle of the jump, and the weight of the bike. Here are data about the maximum height for various bike weights. Plot (weight, height). If the data are linear, draw a trend or best-fit line.

| Weight (pounds) | Height (inches) |
|---|---|
| 19.0 | 10.35 |
| 19.5 | 10.30 |
| 20.0 | 10.25 |
| 20.5 | 10.20 |
| 21.0 | 10.10 |
| 22.0 | 9.85 |
| 22.5 | 9.80 |
| 23.0 | 9.79 |
| 23.5 | 9.70 |
| 24.0 | 9.60 |

2. Is there positive, negative, or no association between bike weight and jump height? Explain your answer. ______________________________

______________________________

3. As the weight increases, the height ______________________________

______________________________

4. Find the slope or rate of change. What does this mean in words? ______________________________

______________________________

5. Predict the maximum height for a bike that weighs 21.5 pounds if all other factors are held constant. ______________________________

______________________________

SHEET 1

***ACTIVITY 12***
***RELATIONS AMONG NBA STATISTICS***

The table contains data on the LA Lakers basketball team taken from the 16 December 1989 Los Angles Laker–Boston Celtic game. Min. indicates minutes played, FG is field goals, FT is free throws, R is rebounds, A is assists, P is personal fouls, and T is total points.

| Player | Min. | FG-A | FT-A | R | A | P | T |
|---|---|---|---|---|---|---|---|
| Cooper | 32 | 4 - 9 | 2 - 2 | 3 | 3 | 2 | 12 |
| Worthy | 35 | 13 - 19 | 2 - 2 | 5 | 2 | 5 | 28 |
| Green | 43 | 8 - 12 | 9 - 12 | 11 | 0 | 4 | 25 |
| Scott | 37 | 9 - 15 | 2 - 2 | 2 | 4 | 0 | 21 |
| Johnson, E. | 43 | 4 - 12 | 8 - 8 | 6 | 21 | 0 | 16 |
| Mcnamar | 5 | 0 - 1 | 0 - 0 | 2 | 0 | 0 | 0 |
| Drew | 8 | 0 - 1 | 2 - 2 | 0 | 0 | 1 | 2 |
| Divac | 23 | 4 - 9 | 2 - 3 | 4 | 4 | 0 | 10 |
| Woolridge | 14 | 2 - 3 | 1 - 1 | 1 | 0 | 0 | 5 |
| Totals | | 44 - 81 | 28 - 32 | 34 | 30 | 12 | 119 |

1. A field goal is worth two or three points. A free throw is worth one point. How many three-point field goals were made in the game by a Laker and who made them?

2. Each member of your group should plot a different set of data. Plot (points scored, fouls); (field goals attempted, field goals made); (minutes played, points scored); (rebounds, assists). Discuss the association between the variables for each graph. Does the plot appear to form a line? If so, draw in the line and find its equation. Write a paragraph summarizing the conclusions you can make about the plots.

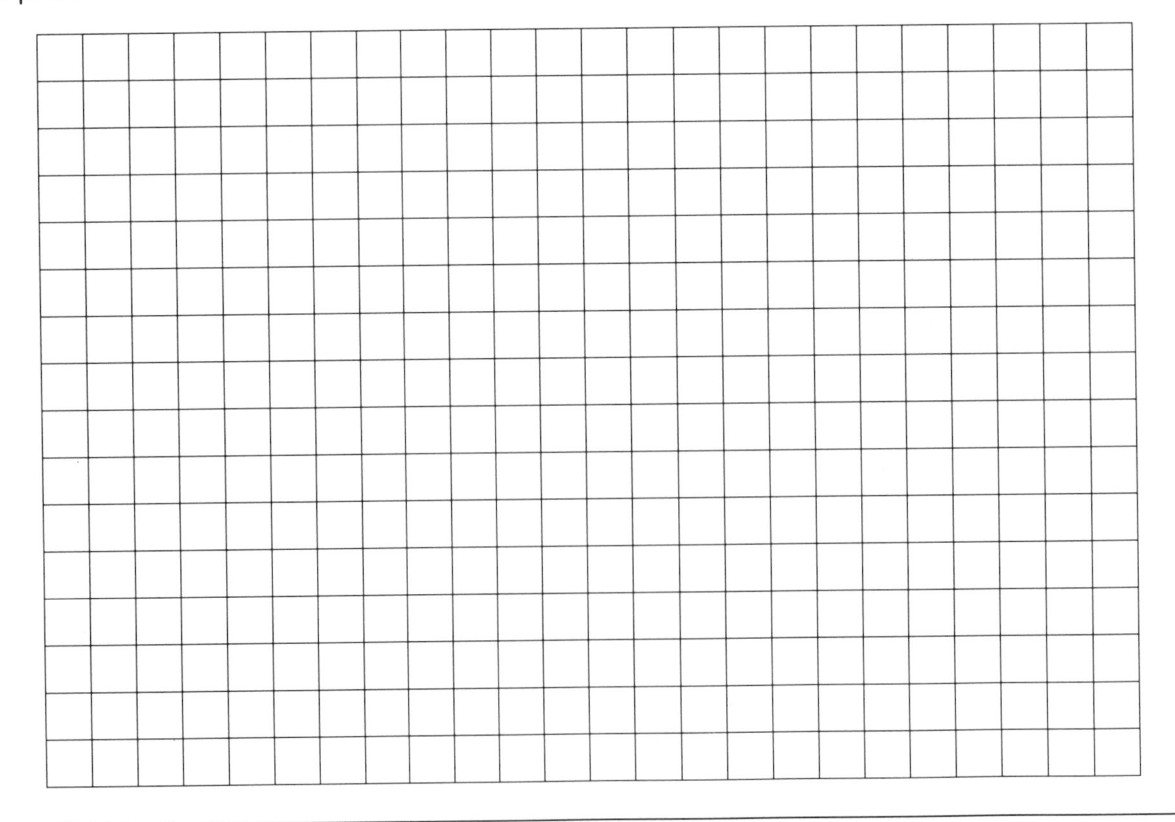

***ACTIVITY 13***
***WEIGHTS AND DRUG DOSES***

SHEET 1

The dosage chart below was prepared by a drug company for doctors who prescribed Tobramycin, a drug that combats serious bacterial infections such as those in the central nervous system, for life-threatening situations.

| Weight (pounds) | Usual dosage (mg) | Maximum dosage (mg) |
|---|---|---|
| 88 | 40 | 66 |
| 99 | 45 | 75 |
| 110 | 50 | 83 |
| 121 | 55 | 91 |
| 132 | 60 | 100 |
| 143 | 65 | 108 |
| 154 | 70 | 116 |
| 165 | 75 | 125 |
| 176 | 80 | 133 |
| 187 | 85 | 141 |
| 198 | 90 | 150 |
| 209 | 95 | 158 |

1. Plot (weight, usual dosage) and draw a best-fit line.
2. Plot (weight, maximum dosage) on the same axes. Draw a best-fit line.
3. Find the slope for each line. What do they mean and how do they compare? ____________________
4. Write the equations of the two lines. ____________________
5. Are the lines parallel? Why or why not? ____________________
6. Use a graphing calculator to plot (usual dosage, maximum dosage). Use the calculator to construct a regression line for this data set. How does this line compare to the two lines found in 1 and 2?

***ACTIVITY 14***
***WINNING TIMES FOR THE WOMEN'S OLYMPIC 400-METER FREESTYLE SWIM***

SHEET 1

The table lists the winning times for the women's 400-meter freestyle swim for the Olympics, 1924–1984.

| Year | Time (min.:sec.) |
|---|---|
| 1924 | 6:02.2 |
| 1928 | 5:42.8 |
| 1932 | 5:28.5 |
| 1936 | 5:26.4 |
| 1948 | 5:17.8 |
| 1952 | 5:12.1 |
| 1956 | 4:54.6 |
| 1960 | 4:50.6 |
| 1964 | 4:43.3 |
| 1968 | 4:31.3 |
| 1972 | 4:19.0 |
| 1976 | 4:09.9 |
| 1980 | 4:08.8 |
| 1984 | 4:07.1 |

Source: *World Almanac and Book of Facts,* 1992 edition, © Pharos Books 1991, New York, N.Y.

1. Using 1920 as the base year, plot (year, time).

2. Construct a best-fit line.

3. What is the slope and what does it mean? ____________________

4. Write the equation of the line. Use the line to predict what the times might have been if the Olympics had been held in 1940 and 1944. ____________________

5. Is it reasonable to use this line to predict the winning time for the 1988 Summer Games? Why or why not? ____________________

6. Look up the winning time for the 400-meter freestyle swim in the 1988 Summer Games and compare it to the time predicted by the best-fit line. ____________________

**Extensions**

7. Use a graphing calculator to find the equation of the regression line.

8. Use a software package such as Data Insights to find the equation of the median-fit line. How do the two equations compare? Which one gives you the "best fit"? Explain your choice.

# CHAPTER 5
# EXPLORING NONLINEAR DATA

The high school standard on functions (NCTM 1989, p. 154) advocates a mathematics curriculum that would enable all students to model real-world phenomena. To implement the standard on statistics (NCTM 1989, p. 167), the curriculum should include experiences that allow students to use curve fitting to predict from data. In addition, it is recommended that the curriculum provide opportunities for college-intending students to transform data to aid in interpretation and prediction. The focus in earlier chapters was on exploring data using simple statistical techniques and on modeling data that appeared to be linear. In this chapter, models for data that are not linear will be developed and analyzed. It is important that students learn to be critical users of statistical tools and be able to interpret the results of a curves fitting transformation intelligently. Questions to be considered include these: Which transformation is appropriate for a given set of data? What does correlation indicate about the data?

*Teaching Matters:* ***It is important that students understand that a strong correlation does not necessarily indicate that a cause-and-effect relationship exists between the variables. Shoe size and reading levels have a high correlation for young children, but big feet do not cause children to read well! They are both functions of another variable—age.***

There are several alternatives to using a linear model. Typically the most suitable models for a high school student who is at the level of second-year algebra are quadratic (or power) and exponential or logarithmic. The following problem situation, which provides the context for the first activity at the end of this chapter, illustrates how the investigation of nonlinear data can be incorporated in the mainline curriculum.

*Try This:* ***Most states have a driver's manual that gives a table similar to the one below expressing the number of feet required for a car to stop and the speed at which it is traveling prior to applying the brakes. The manuals also give the "rule of thumb" of following two car lengths for every ten miles per hour. Does the linear "rule of thumb" fit the data in the chart?***

| Speed (mph) | Stopping distance (ft.) |
|---|---|
| 10 | 15 |
| 20 | 40 |
| 30 | 75 |
| 40 | 120 |
| 50 | 175 |
| 60 | 240 |
| 70 | 315 |
| 80 | 400 |

### THE PROBLEM

A forester would like to know how large a tree will be if it grows for fifty years. Should the tree be cut when it is thirty-five years old, or will the growth of the tree over an additional fifteen-year span significantly increase the size of the tree? How can the diameter of a tree be predicted from its age? The data in table 5.1, from *Exploring Data* (Landwehr and Watkins 1986), give the age and diameters at chest height of twenty-seven chestnut oak trees planted on a relatively poor site.

Table 5.1

| Age (years) | Diameter (inches) | Age (years) | Diameter (inches) |
|---|---|---|---|
| 4 | 0.8 | 23 | 4.7 |
| 5 | 0.8 | 25 | 6.5 |
| 8 | 1.0 | 28 | 6.0 |
| 8 | 2.0 | 29 | 4.5 |
| 8 | 3.0 | 30 | 6.0 |
| 10 | 2.0 | 30 | 7.0 |
| 10 | 3.5 | 33 | 8.0 |
| 12 | 4.9 | 34 | 6.5 |
| 13 | 3.5 | 35 | 7.0 |
| 14 | 2.5 | 38 | 5.0 |
| 16 | 4.5 | 38 | 7.0 |
| 18 | 4.6 | 40 | 7.5 |
| 20 | 5.5 | 42 | 7.5 |
| 22 | 5.8 | | |

Source: Chapman and Demeritt, *Elements of Forest Mensuration*

The first step is to look at the data to determine what they mean and where they came from. Then plot the points and look for a relation or pattern. The scatterplot in figure 5.1 looks somewhat linear, and a straight line might be an efficient way to characterize the relationship. Figure 5.1 shows a median-fit line through the data points. A careful look at the data points in relation to the line reveals several disturbing features. The points at the beginning and end of the data are below the line, and the points in the middle are above the line. This indicates that a straight line does not fit the data well. What does this suggest about the rate of change of diameter (or slope) in terms of age? For every year increase in age, the growth in diameter is not constant; the relation is not linear. By inspecting the data points, it seems as if the growth rate of the diameter increases rapidly at first and slows down as the tree ages.

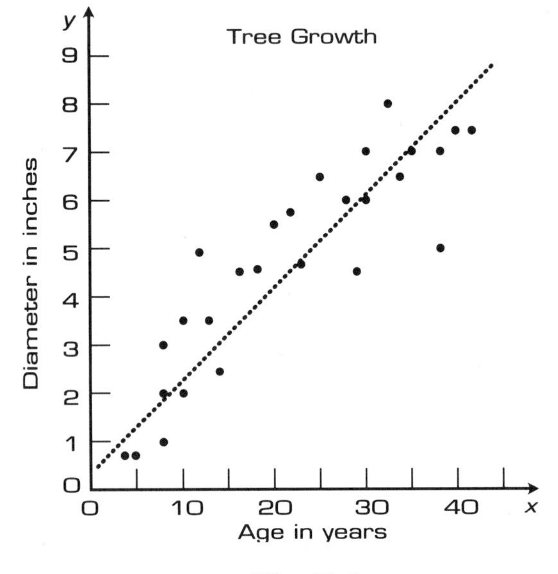


***Fig. 5.1***

What kind of relation *does* exist between the age of a tree and its diameter? It is possible to draw in a curve by hand that seems to fit the data. For some students, this may be sufficient; however, to summarize the data, the curve must be quantified. That is, an equation is needed. The task is to find a function to describe the pattern formed by the points. Students could guess at a relation and experiment with different combinations of rules to find one that seems to be a "best fit," but this is tedious, time consuming, and seldom accurate. Since it is assumed that students already know how to work with straight lines, is it possible to make the data linear? A median-fit line could then be used to describe the relation symbolically. Perhaps the data could be transformed in some fashion. One interesting mathematical concept is that of logarithms, the language of exponents. The log of a product becomes a sum; using logarithms to transform the data may have the effect of making the data linear.

*Teaching Matters:* ***This activity builds on the concept of the median-fit line, which could be covered in first-year algebra. Better students should also apply the same process to a regression line and should recognize that a regression line is a function of the mean and has the same sensitivity to extreme values.***

Assign the following tasks to groups of students after they have analyzed the original graph of (age, diameter).

1. Take the natural logarithm of the diameter and graph (age, ln diameter).
2. Take the common logarithm of the diameter and graph (age, log diameter).
3. Take the natural logarithm of age and graph (ln age, diameter).
4. Take the common logarithm of age and graph (log age, diameter).

5. Take the natural logarithm of both age and diameter and graph (ln age, ln diameter).
6. Take the common logarithm of both the age and the diameter and graph (log age, log diameter).

Make sure students use graph paper and have reasonable scales or use a graphing calculator or computer software to create an accurate graph. A good tool might be the software Data Models (Sunburst 1991), which will transform the data and then display and graph the transformed data. After the points have been plotted, have students construct a median-fit line and analyze the relation of the points to the line. Has the transformation "straightened" the data at all? Each team should report their results to the class. Table 5.2 contains the original and the transformed data.

***Try This:* Have students investigate the similarity between the natural and common logarithmic transformations by analyzing the graphs and the scales. One is a scalar multiple of the other. Is distance preserved under such a transformation?**

Table 5.2

| Age in years | Diameter in inches | ln (age) | log (age) | log (diameter) | ln (diameter) |
|---|---|---|---|---|---|
| 4 | 0.8 | 1.3863 | 0.6021 | −0.0969 | −0.2231 |
| 5 | 0.8 | 1.6094 | 0.6990 | −0.0969 | −0.2231 |
| 8 | 1 | 2.0794 | 0.9031 | 0.0000 | 0.0000 |
| 8 | 2 | 2.0794 | 0.9031 | 0.3010 | 0.6931 |
| 8 | 3 | 2.0794 | 0.9031 | 0.4771 | 1.0986 |
| 10 | 2 | 2.3026 | 1.0000 | 0.3010 | 0.6931 |
| 10 | 3.5 | 2.3026 | 1.0000 | 0.5441 | 1.2528 |
| 12 | 4.9 | 2.4849 | 1.0792 | 0.6902 | 1.5892 |
| 13 | 3.5 | 2.5649 | 1.1139 | 0.5441 | 1.2528 |
| 14 | 2.5 | 2.6391 | 1.1461 | 0.3979 | 0.9163 |
| 16 | 4.5 | 2.7726 | 1.2041 | 0.6532 | 1.5041 |
| 18 | 4.6 | 2.8904 | 1.2553 | 0.6628 | 1.5261 |
| 20 | 5.5 | 2.9957 | 1.3010 | 0.7404 | 1.7047 |
| 22 | 5.8 | 3.0910 | 1.3424 | 0.7634 | 1.7579 |
| 23 | 4.7 | 3.1355 | 1.3617 | 0.6721 | 1.5476 |
| 25 | 6.5 | 3.2189 | 1.3979 | 0.8129 | 1.8718 |
| 28 | 6.0 | 3.3322 | 1.4472 | 0.7782 | 1.7918 |
| 29 | 4.5 | 3.3673 | 1.4624 | 0.6532 | 1.5041 |
| 30 | 6.0 | 3.4012 | 1.4771 | 0.7782 | 1.7918 |
| 30 | 7.0 | 3.4012 | 1.4771 | 0.8451 | 1.9459 |
| 33 | 8.0 | 3.4965 | 1.5185 | 0.9031 | 2.0794 |
| 34 | 6.5 | 3.5264 | 1.5315 | 0.8129 | 1.8718 |
| 35 | 7.0 | 3.5553 | 1.5441 | 0.8451 | 1.9459 |
| 38 | 5.0 | 3.6376 | 1.5798 | 0.6990 | 1.6094 |
| 38 | 7.0 | 3.6376 | 1.5798 | 0.8451 | 1.9459 |
| 40 | 7.5 | 3.6889 | 1.6021 | 0.8751 | 2.0149 |
| 42 | 7.5 | 3.7377 | 1.6232 | 0.8751 | 2.0149 |

Figures 5.2–5.7 display the transformed data and median-fit lines corresponding to transformations numbered 1–6 respectively. Which of the six approaches produces points that seem to be best represented by a straight line? As seen in figures 5.2 and 5.3, taking either the natural logarithm or the common logarithm of the diameter seemed to accent the original curve. As seen in figures 5.4, 5.5, 5.6, and 5.7, taking the natural or common logarithm of both variables produced a relatively straight line, as did taking the logarithm of just the age. Using common

logarithms, however, did not give a very large spread for the range. Looking closely, we see that the difference between the natural and common logarithm transformations seems to be only one of scale. Because a larger spread might make the pattern easier to work with and reduce some of the deviation, it seems reasonable to concentrate on transformations using the natural logarithm. Both figures 5.4 and 5.6 seem to give relatively good fits between the line and the data. Note that in figure 5.6, however, the points are far from the line for small $x$'s and close to the line for large $x$'s. This means that if the line is used to predict for a small $x$, the variation (or "error") will be large, much larger than the error for bigger domain values. Because of the inconsistency in the size of the error, taking the natural logarithm of only the age seems to be the best choice for transforming the data.

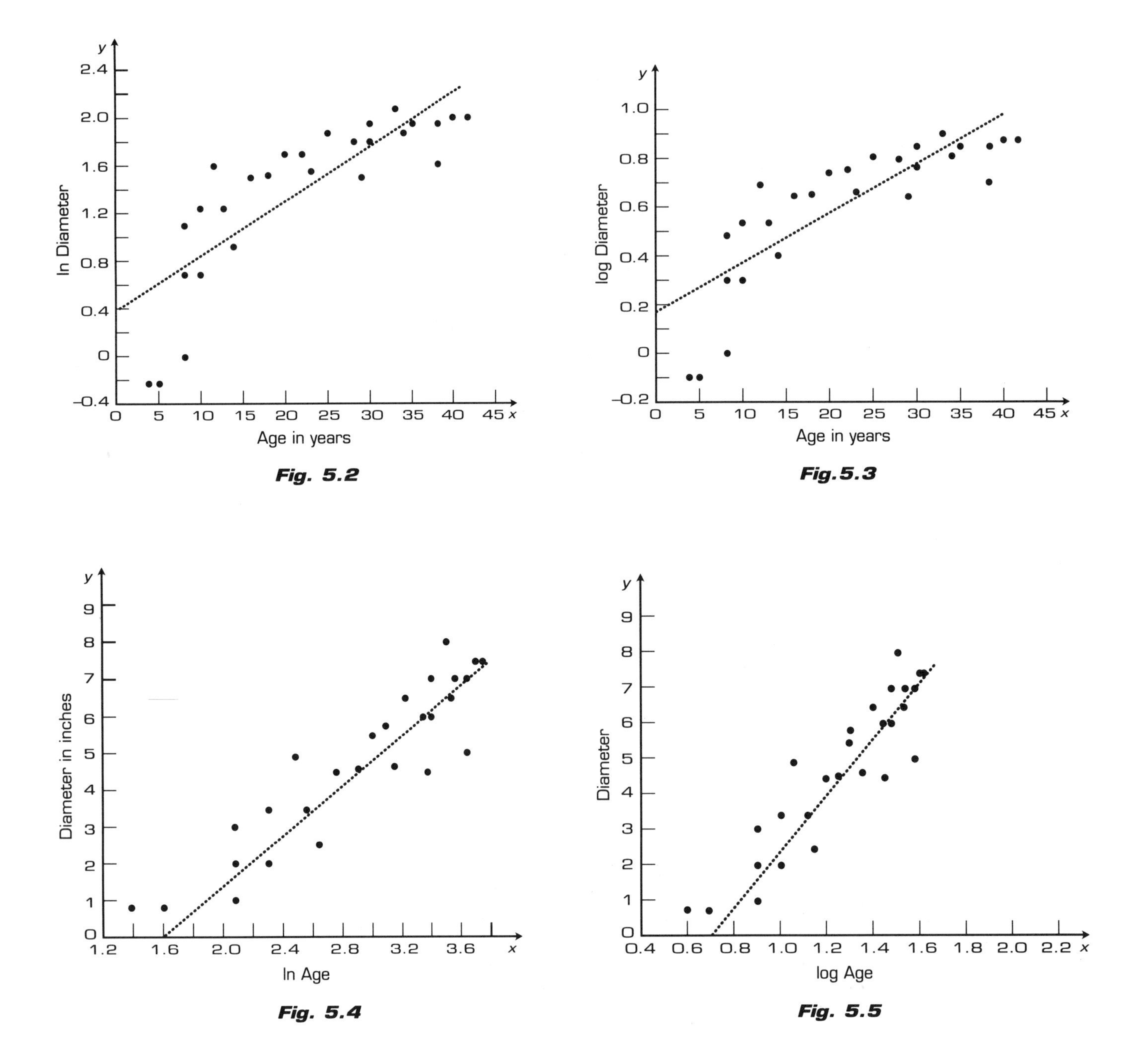


Fig. 5.2

Fig. 5.3

Fig. 5.4

Fig. 5.5

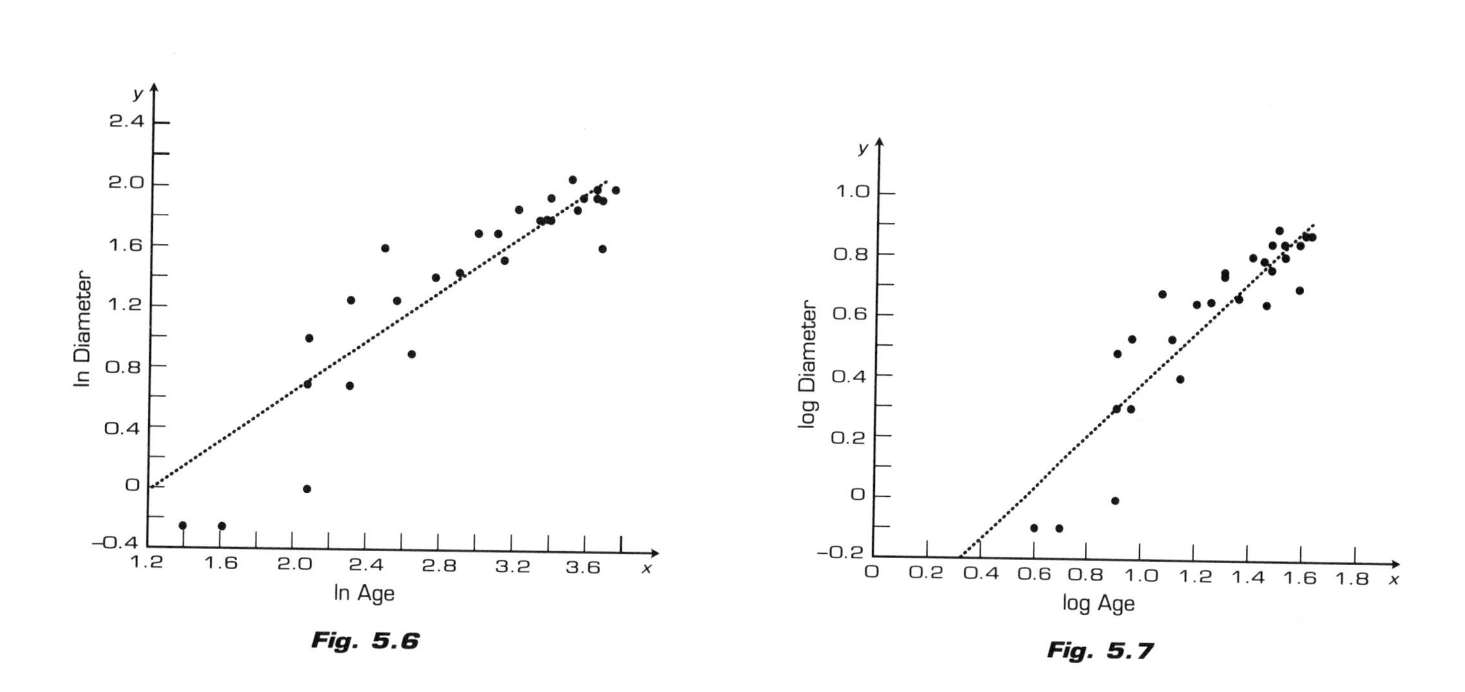


***Fig. 5.6***

***Fig. 5.7***

The graph in figure 5.4 indicates that the relation between the points has been "straightened" by the log transformation. The median-fit line seems to summarize the relation between the natural logarithm of the age of a tree and its diameter, and from this, the relation between the age of the tree and its diameter.

*Try This:* ***Have students plot the residuals for the different methods of data transformation and analyze how this will help determine which transformation most adequately summarizes the data.***

To find the equation of the line, select any two points on the median-fit line. For example, for the points (2.1, 1.8) and (2.6, 3.5), the slope of the line is

$$\frac{3.5 - 1.8}{2.6 - 2.1} = 3.4.$$

Using the point-slope form of the equation of a line, we find that $y - y_1 = m(x - x_1)$, and hence $y - 1.8 = 3.4(x - 2.1)$, or $y = 3.4x - 5.3$.

Remember that "$x$" is really the natural log of the age.

Write the equation: $y = 3.4 \ln x - 5.3$.

Just as with exploring linear data, the result can be used to predict.

If the tree is thirty-six years old, what is an estimate for the diameter?

$d = 3.4 \ln 36 - 5.3$ ("$y$" represents the diameter)

$\approx 6.8$ inches

Thus, a tree that is thirty-six years old has a diameter that is approximately 6.8 inches.

#### *HOW DO YOU KNOW IF YOU HAVE A GOOD RULE FOR THE PATTERN?*

Is 6.8 inches really a good estimate? How could you tell? One approach is to find some process to calculate an "average" error for the difference between the actual $y$-value and the predicted $y$-value by using the equation just generated. Some of the differences will be positive, and some will be negative. The statistical term for the error is *residual:* the vertical dis-

*Teaching Matters:* ***Analogies are often helpful when introducing new content. The mean squared error is similar to variance, and the root mean error is similar to the standard deviation. They both represent an "average" distance from a given reference.***

tance above (+) or below (–) the line of fit. To avoid working with both positive and negative numbers, mathematicians usually take the absolute value or square the differences. For example, to find the mean squared difference, consider the data point (12, 4.9). For a twelve-year-old tree, the actual diameter was 4.9 inches. The equation generated from the median-fit line predicts that a twelve-year-old tree will have a diameter of 3.1 inches. The difference between the actual *y*-value (4.9) and the predicted *y*-value (3.1) is 1.8, which when squared is 3.24 . For a thirty-four-year-old tree, the actual diameter was 6.5 inches, and the predicted diameter was 6.6. The difference between the two values, –0.1, when squared is 0.01. To find the mean, the squared differences are added, and the result is divided by the number of data points. The mean squared error for predicting the diameter of trees from their ages using the median-fit equation is calculated in table 5.3 and equals approximately 0.8188. To put this number in the same scale as the data, take

Table 5.3
Equation used: $y = 3.37 \ln x - 5.26$

| Age | Diameter | Predicted diameter | Difference | Difference squared |
|---|---|---|---|---|
| 4 | 0.8 | –0.58818800302 | 1.3881880030 | 1.9270659317 |
| 5 | 0.8 | 0.163805764902 | 0.6361942350 | 0.4047431047 |
| 8 | 1.0 | 1.747717995461 | –0.747717995 | 0.5590822007 |
| 8 | 2.0 | 1.747717995461 | 0.2522820045 | 0.0636462098 |
| 8 | 3.0 | 1.747717995461 | 1.2522820045 | 1.5682102189 |
| 10 | 2.0 | 2.499711763390 | 0.499711763 | 0.2497118464 |
| 10 | 3.5 | 2.499711763390 | 1.0002882366 | 1.0005765563 |
| 12 | 4.9 | 3.114135409786 | 1.7858645902 | 3.1893123346 |
| 13 | 3.5 | 3.383879334645 | 0.1161206653 | 0.0134840089 |
| 14 | 2.5 | 3.633623200803 | –1.133623201 | 1.2851015614 |
| 16 | 4.5 | 4.083623993948 | 0.4163760060 | 0.1733689784 |
| 18 | 4.6 | 4.480552824110 | 0.1194471758 | 0.0142676278 |
| 20 | 5.5 | 4.835617761877 | 0.6643822381 | 0.4414037583 |
| 22 | 5.8 | 5.156813067818 | 0.6431869321 | 0.4136894297 |
| 23 | 4.7 | 5.306615507681 | –0.606615507 | 0.3679823741 |
| 25 | 6.5 | 5.587611529806 | 0.9123884701 | 0.8324527205 |
| 28 | 6.0 | 5.969529199290 | 0.0304708007 | 9.2846970E–4 |
| 29 | 4.5 | 6.087786947054 | –1.587786947 | 2.5210673892 |
| 30 | 6.0 | 6.202035176201 | –0.202035176 | 0.0408182124 |
| 30 | 7.0 | 6.202035176201 | 0.7979648237 | 0.6367478600 |
| 33 | 8.0 | 6.523230482142 | 1.4767695179 | 2.1808482089 |
| 34 | 6.5 | 6.623834967956 | –0.123834967 | 0.0153350992 |
| 35 | 7.0 | 6.721522967219 | 0.2784770327 | 0.0775494577 |
| 38 | 5.0 | 6.998665358278 | –1.998665358 | 3.9946632144 |
| 38 | 7.0 | 6.998665358278 | –0.0013346417 | 1.7812685E–6 |
| 40 | 7.5 | 7.171523760364 | 0.3284762396 | 0.1078966400 |
| 42 | 7.5 | 7.335946613615 | 0.1640533863 | 0.0269135135 |

Sum = 22.107

Mean Squared Error = 22.107 ÷ 27 = 0.8188*

Root Mean Squared Error = $\sqrt{0.8188} \approx 0.905$

*Rounding errors may account for slightly different results.

the square root ($\sqrt{.8188} \approx 0.905$). This root mean square error indicates on the average how far above or below the median-fit line the points lie. A prediction for the diameter of the thirty-six-year-old tree would be 6.9 inches plus or minus approximately 0.9 inches, or from 6.0 inches to 7.8 inches. (To calculate a statistic where *n*, the number of data points, is a divisor under certain conditions, divide by $n - 1$; under other other conditions, divide by $n - 2$, and so on. To maintain the notion of average, do all the division by *n*.)

### THE CORRELATION COEFFICIENT

Another way to test how well the median-fit line can be used to predict is to calculate a statistic called the correlation coefficient, *r*. If the relation is linear, the correlation coefficient measures the strength of the relationship between the two variables. Basically, *r* is a function of the difference of each *x*-value from the mean *x*, each *y*-value from the mean *y*, and the standard deviations of the *x*- and *y*-values. The products of the differences are summed; if the *x*- and *y*-values are either both below or both above the mean, the product is positive. If this is true for most of the points, the sum is positive. If the *x*-value is above and the *y*-value below their respective means or vice versa, the product is negative. If this is true for most of the points, the sum is negative. If both situations exist for approximately equal sets of data points. the sum will be 0. To scale and "average" the sum, the result is divided by the standard deviations for each variable and by 1 less than the number of data points (Freedman, Pisani, and Purves 1980).

*Teaching Matters:* **Asking students to explain in words formulas such as that for the correlation coefficient**

$$r = \frac{\sum_{i=1}^{r} (\bar{x} - x_i)(\bar{y} - y_i)}{(n-1)s_x s_y}$$

**helps foster the development of symbol sense.**

Most calculators with a statistics mode have an "r" key that will produce this number after the data are entered, and some computer software will do so also. As a consequence of the products in the definition, if *r* is close to 0, there is very little association between the variables (and the relation between the points is not linear). Remember that this means that some small *x*-values correspond to large *y*-values and others to small *y*-values; that is, some trees with small diameters would be old and some would be young. If *r* is close to 1, there is a strong positive association: trees with small diameters tend to be young trees. If *r* is close to −1, there is a strong negative association: trees with small diameters would tend to be old trees. A geometric interpretation of *r* is given in figure 5.8.

*Try This:* **Have students estimate the correlation between two data sets from their scatterplot. Have students draw a scatterplot that might represent correlations of 0, −0.5, 0.2, 0.5, and 0.8.**

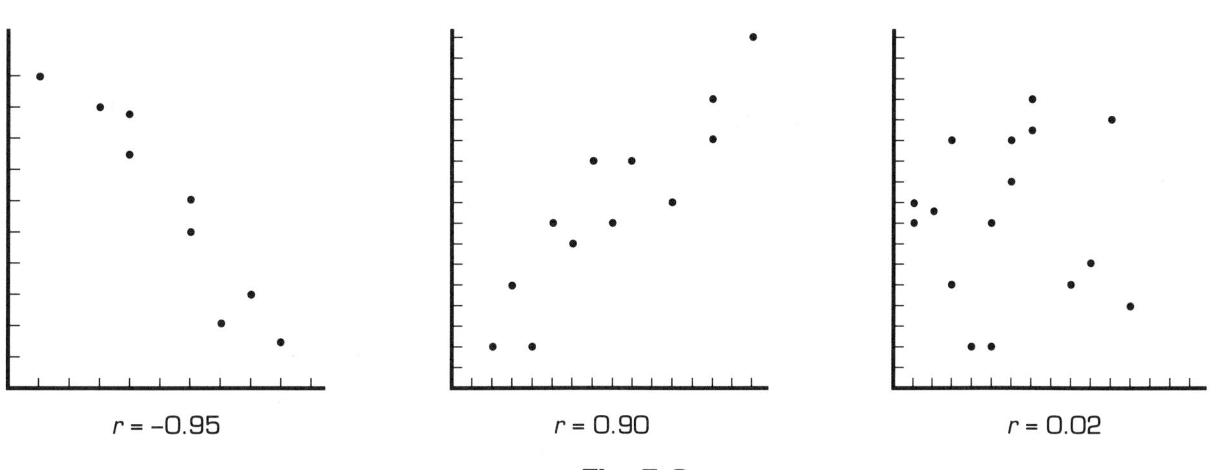


**Fig. 5.8**

*Try This:* ***Have students use their calculators to obtain the regression equation for the data, graph that equation on the same axes as the median-fit line, and determine which of the two lines is the best summary for the data. Be sure they can justify their choice of line.***

The relation between the natural logarithm of the age of a tree and its diameter has a correlation coefficient of 0.92, which indicates a strong positive association. Without transforming the data, $r$ was 0.88; the association was good, which is often true for a small section of the domain. The logarithmic transformation, however, increased the correlation and also decreased our error.

The $r$ for figure 5.6 was 0.91. This is very close to the correlation for figure 5.4, but the graph for figure 5.6 has already indicated this is not as good a representation for the data as the line in figure 5.4. *Without the graph it would be very difficult to determine which was the better fit.* To attempt to make conclusions about data without looking at a graph of the data can be misleading and lead to serious errors in judgment. The mean squared error for figure 5.6 was 0.98, which is not far from 0.90. In both cases the error is close to 1.0, but remember this is an average. As indicated earlier, the error for small values of $x$ will be large, and the error for large values of $x$ will be small. When the mean is used as a measure of center, knowledge of the data is important, and caution must be exercised in interpreting the results.

### A MATHEMATICAL INTERPRETATION

Essentially such reexpression is simply an application of the composition of functions. The equation for the median-fit line in figure 5.4 is of the form $y = m\ln(x) + b$. Using properties of logarithms, we find that this is equivalent to $y = \ln x^m + b$, a logarithmic function. If the data set $(x, y)$ can be represented by an unknown function $f$ and is reexpressed as $(\ln(x), y)$, then the composition is $f(g(x))$, where function $g = \ln x$, the natural log function. If $f(g(x))$ is linear, then $f(x) = m\,g(x) + b = m\ln(x) + b$.

If the transformation obtained by taking the logarithm of the diameter had given a best fit, the equation of the median-fit line would have been $\ln y = mx + b$. This is equivalent to $y = e^{mx+b}$, the exponential function. Then the reexpressed data set $(x, \ln(y))$ is simply $g(f(x))$, where the function $g$ is the natural log function. Quite literally, the $y$-values of function $f$ are the input value for $g$. If the result is a line, that is, $g(f(x)) = mx + b$, then the unknown function can be found by solving for $f$: $f(x) = g^{-1}(mx + b) = \exp(mx + b) = e^{mx+b}$, or $e^b e^{mx}$.

If the transformation in figure 5.6 had given the best fit, the equation would have been of the form $\ln y = m\ln(x) + b$. Again, using properties of logarithms, we find that this is equivalent to $\ln y = \ln x^m + b$. For some $c$, $b = \ln c$, or $e^b = c$, so $\ln y = \ln x^m + \ln c$. (If $b$ is negative, use the division property.) Again using properties of logarithms, we find that $\ln y = \ln cx^m$ or $y = cx^m$ which is called a power function. In terms of composite functions, when the data set $(x, y)$ is reexpressed as $(\ln(x), \ln(xy))$, the result is a composition of three functions, $h(f(g(x)))$. If this is a line, then $f(x) = h^{-1}(mg(x) + b) = \exp(m\ln(x) + b) = e^{m\ln x}e^b = e^b x^m$. If $m$ is $-1$, the function is hyperbolic; if $m$ is 2, the function is quadratic; and so on. Note that the slope of the median-fit line determines the power of $x$ in the relation.

Students who have studied polynomial, logarithmic, and exponential functions will recognize how the plots in the diagrams are related to those functions. Encourage students to make these connections, and as they become familiar with the process, use this knowledge to simplify finding an appropriate transformation. It is important, however, to recognize that students need hands-on experience with the various transformations. Students should also be aware that there are many transformations other than those discussed here (cf. NCCSM 1988).

Eventually they will recognize the mathematics inherent in the problem that will point the way to an appropriate transformation. For example, the relation between height and weight might be cubic because height is linear and weight is associated with volume, which is three dimensional. Software packages such as Data Models or Cricket Graph and graphing calculators such as the Casio fx 7700G or the Texas Instuments TI-81 are powerful and easy-to-use tools for investigating nonlinear data. Specific activities featuring the TI-81 can be found in Burrill and Hopfensperger (1992).

The three classroom-ready activities that follow illustrate how explorations of nonlinear data can be woven into the high school mathematics curriculum. In addition to introducing students to powerful methods of data analysis, these activities provide interesting contexts for applications of algebra.

## ACTIVITY 15
## TREE GROWTH

SHEET 1

How can the diameter of a tree be predicted from its age? (The data below were collected from chestnut oak trees growing in a relatively poor site.)

1. Find the common and natural logarithms for each piece of data.

| Age (yrs.) | Diameter (in.) | log Age | log Diameter | ln Age | ln Diameter |
|---|---|---|---|---|---|
| 4 | 0.8 | | | | |
| 5 | 0.8 | | | | |
| 8 | 1.0 | | | | |
| 8 | 2.0 | | | | |
| 8 | 3.0 | | | | |
| 10 | 2.0 | | | | |
| 10 | 3.5 | | | | |
| 12 | 4.9 | | | | |
| 13 | 3.5 | | | | |
| 14 | 2.5 | | | | |
| 16 | 4.5 | | | | |
| 18 | 4.6 | | | | |
| 20 | 5.5 | | | | |
| 22 | 5.8 | | | | |
| 23 | 4.7 | | | | |
| 25 | 6.5 | | | | |
| 28 | 6.0 | | | | |
| 29 | 4.5 | | | | |
| 30 | 6.0 | | | | |
| 30 | 7.0 | | | | |
| 33 | 8.0 | | | | |
| 34 | 6.5 | | | | |
| 35 | 7.0 | | | | |
| 38 | 5.0 | | | | |
| 38 | 7.0 | | | | |
| 40 | 7.5 | | | | |
| 42 | 7.5 | | | | |

2. Working in groups and using rectangular graph paper, make the following scatterplots. Construct a median-fit line for your points.
   a. (Age, Diameter)
   b. (Age, log Diameter)
   c. (Age, ln Diameter)
   d. (log Age, Diameter)
   e. (ln Age, Diameter)
   f. (ln Age, ln Diameter)
   g. (log Age, log Diameter)

***ACTIVITY 15***
***TREE GROWTH (CONTINUED)*** SHEET 2

3. The median-fit line is a good way to summarize the data if most of the points are close to the line and the points do not look "curved" around the line. How well does a median-fit line represent the data sets in part 2? Justify your answer.

   a. (Age, Diameter) ____________________

   b. (Age, log Diameter) ____________________

   c. (Age, ln Diameter) ____________________

   d. (log Age, Diameter) ____________________

   e. (ln Age, Diameter) ____________________

   f. (ln Age, ln Diameter) ____________________

   g. (log Age, log Diameter) ____________________

4. Share your results with the rest of the class. Which transformation "straightens" the data the most? Which transformation makes the median-fit line a good fit for the data? Justify your answer.

5. Using the graph for the transformed data that has the "best" fit for the points, select two points on the median-fit line to find the equation of the line.

6. If your answer is not a function of *y*, use what you know about logarithms to write the equation of the line without logarithms. Use a graphing calculator or graphing software to graph this function. How well does it relate to your original set of data?

7. Use your equation to predict the diameter of the tree when it is 36 years old and when it is 50 years old.

8. Suppose you cut down on that site a tree with a diameter of 8.5 inches. What is an estimate for the age of the tree?

## *ACTIVITY 15*
## *TREE GROWTH (CONTINUED)*

SHEET 3

9. Use your equation to find the mean squared error by filling out the chart.

| Age (yrs.) | Diameter (in.) | Predicted diameter | Difference | Difference squared |
|---|---|---|---|---|
| 4 | 0.8 | | | |
| 5 | 0.8 | | | |
| 8 | 1.0 | | | |
| 8 | 2.0 | | | |
| 8 | 3.0 | | | |
| 10 | 2.0 | | | |
| 10 | 3.5 | | | |
| 12 | 4.9 | | | |
| 13 | 3.5 | | | |
| 14 | 2.5 | | | |
| 16 | 4.5 | | | |
| 18 | 4.6 | | | |
| 20 | 5.5 | | | |
| 22 | 5.8 | | | |
| 23 | 4.7 | | | |
| 25 | 6.5 | | | |
| 28 | 6.0 | | | |
| 29 | 4.5 | | | |
| 30 | 6.0 | | | |
| 30 | 7.0 | | | |
| 33 | 8.0 | | | |
| 34 | 6.5 | | | |
| 35 | 7.0 | | | |
| 38 | 5.0 | | | |
| 38 | 7.0 | | | |
| 40 | 7.5 | | | |
| 42 | 7.5 | | | |

The sum of the squares is ____________

The mean squared error $= \dfrac{\textit{sum of squares}}{\textit{number of data points}} =$ ____________ $=$

The square root of the mean squared error $= \sqrt{\quad} =$

10. Find the root mean squared error for the median-fit line for (Age, Diameter). How does this compare to the error for the line (ln Age, Diameter)?

11. Find the correlation coefficient for your line. What does this indicate?

12. What would the plot look like if the correlation were 0.3? What would this mean in terms of the age and diameter of trees?

## ACTIVITY 16
## MUSTANG PRICES

SHEET 1

1. These prices of used Ford Mustangs were taken from the *Milwaukee Journal* of 28 January 1990. Plot (Age, Price).

| Year | Age (yrs.) | Price |
|---|---|---|
| 1982 | 8 | $1 400 |
| 1983 | 7 | 2 595 |
| 1986 | 4 | 4 488 |
| 1986 | 4 | 5 495 |
| 1986 | 4 | 4 995 |
| 1987 | 3 | 6 000 |
| 1989 | 1 | 12 895 |
| 1979 | 11 | 1 350 |
| 1980 | 10 | 950 |
| 1985 | 5 | 1 900 |
| 1987 | 3 | 5 890 |
| 1988 | 2 | 6 850 |

2. Are the data linear? If so, find the median-fit line and write its equation.

   What would the slope represent? The intercepts?

3. If the data are not linear, find a transformation that will straighten the data so a median-fit line will summarize the data fairly well. Write the equation of the median-fit line.

4. Find the correlation coefficient for the data. What does this indicate about the association between price and age of the car?

5. On the basis of your data and your equation, for how much would you sell a six-year-old Mustang?

6. If you had $4000 to spend on a used Mustang, how old a car could you expect to buy?

SHEET 1

***ACTIVITY 17***
***QUALIFYING TIMES***

The qualifying times for the Women's A Division 100-yard Butterfly Swim in the YWCA 1987 Championships are given in the table below.

| Age | Time | Time (sec.) |
|---|---|---|
| 22 | 1:29 | 89 |
| 27 | 1:31 | 91 |
| 32 | 1:37 | 97 |
| 42 | 1:53 | 113 |
| 47 | 2:01 | 121 |
| 52 | 2:09 | 129 |
| 57 | 2:21 | 141 |
| 62 | 2:36 | 156 |
| 67 | 2:40 | 160 |
| 72 | 2:45 | 165 |

1. Plot (Age, Time). Does the plot of the data appear linear?

2. Find a transformation that will straighten the data and find the median-fit line for the transformed data.

3. Use your results to predict the qualifying time for a thirty-seven-year-old woman. How close was this to 105 seconds, the actual qualifying time?

4. Find the root mean squared error for your line. What does this mean?

5. Use an appropriate graphing calculator to transform the data. Find the equation and the correlation coefficient for each set of transformed data. Graph the equations and the data. Which equation seems to be the best "fit"?

   Linear transformation: ____________ Correlation: ____________

   Logarithmic transformation: ____________ Correlation: ____________

   Exponential transformation: ____________ Correlation: ____________

   Power transformation: ____________ Correlation: ____________

6. What could explain any difference between the results produced by your first transformation and by using the calculator?

# CHAPTER 6
# CHI SQUARE: A MEASURE OF DIFFERENCE

The *Curriculum and Evaluation Standards* (NCTM 1989) recommends that designing, conducting, and interpreting experiments should be an integral part of every student's experiences in statistics. In addition, students should understand sampling and recognize its role in statistical claims as well as begin to appreciate ideas involved in hypothesis testing. In chapter 1, statistical inference was discussed from two perspectives: comparing a sample to a known population or standard behavior and using a known sample to make some statements about a population. In the first case, a statistic calculated about a sample is compared to a known statistic such as a sample mean to a population mean. This chapter draws on material from *Using Statistics* (Travers et al. 1985) to explore the chi square statistic by generating data from classroom experiments and provides students with one method for making a decision about an event that does not seem to conform to an expected behavior.

Consider tossing a six-sided die sixty times. You would expect to get a given face "about" ten times. But suppose that the die had a chipped edge, and your results were those in table 6.1. There is a considerable difference between what you would expect to happen and what actually occurred. Was this because of the chipped edge, or could this variation have happened naturally? Is there something interesting in the die itself (the chipped edge) that will prevent conformity, or is the deviation from the expected outcome insignificant?

**Teaching Matters: *Discuss how one would know a die to be fair before experimenting with it. (In fact, we really would not.) The mathematical assumption of fairness leads to these theoretical probabilities for a chi square distribution, but high school students will profit from an intuitive approach as described here.***

Table 6.1

| Face | Tally | Observed results |
|---|---|---|
| 1 | /////// | 7 |
| 2 | ///// | 5 |
| 3 | ///////// | 9 |
| 4 | /////////// | 11 |
| 5 | /////////////// | 15 |
| 6 | ///////////// | 13 |
| | | 60 |

One way to find out is to continue tossing the die and recording the results for many tosses. After a very large number of tosses, the experimental probability of an event should be almost the same as the theoretical. Statisticians, however, do not care to run the experiment so many times and have come up with another method to find out if the die was fair. A number called the *chi square* statistic is calculated to measure the difference between what actually happened and what was expected to happen. The difference can be explained in two ways: either the difference is due to chance (random variation) or it is due to some cause inherent in the object being analyzed.

**Try This: *Investigate a spherical die to see if it is fair.***

### HOW TO CALCULATE CHI SQUARE

Record the observed results and the expected results in a table with headings similar to those in table 6.2. Because chi square is a measure of the difference between the observed results and the expected results, each observed value is subtracted from the number expected. To elimi-

*Teaching Matters:* ***This version of chi square only compares observed frequencies to theoretical frequencies. This is not the same as looking at a two-way table to determine if the patterns are the same.***

*Teaching Matters:* ***It is important that students understand that a chi square depends on the number of outcomes of the experiment. If an experiment has four outcomes, a new chi square distribution would have to be generated to use as a standard.***

*Assessment Matters:* ***Students, especially those who are younger or less mathematically able, will need to experiment with random-number generators to get a "feel" for the meaning of the chi square test. As they work, observe them carefully to determine how well they understand.***

*Teaching Matters:* ***After students have computed and tested chi squares in several situations, have them estimate from a given data table whether or not a chi square would be large or small. Discuss the meaning of large and small.***

*Try This:* ***Have students investigate this question: Is there a maximum chi square for a given number of outcomes and sample size?***

nate the negative values, square the difference. To avoid inflating the difference by the number of trials, divide each difference by the expected result. The sum of all these quotients is a number called "chi square." (The division is performed before adding because the expected results need not be equally likely.) The symbol for chi square is $\chi^2$. Table 6.2 shows the chi square statistic calculated for tossing the chipped die.

Table 6.2

| Face | Observed | Expected | O – E | $(O-E)^2$ | $\frac{(O-E)^2}{E}$ |
|---|---|---|---|---|---|
| 1 | 7 | 10 | –3 | 9 | 0.9 |
| 2 | 5 | 10 | –5 | 25 | 2.5 |
| 3 | 9 | 10 | –1 | 1 | 0.1 |
| 4 | 11 | 10 | 1 | 1 | 0.1 |
| 5 | 15 | 10 | 5 | 25 | 2.5 |
| 6 | 13 | 10 | 3 | 9 | 0.9 |
| Sum | 60 | 60 | | | $\chi^2 = 7.0$ |

***HOW TO INTERPRET A CHI SQUARE***

For the data in table 6.2, $\chi^2 = 7$. What does this mean? How does this explain the variation? If what is expected to happen does occur, what would be the value for $\chi^2$? When there is little or no difference between the observed and expected outcomes, the numerator is close to 0. Then $\chi^2$ is very small, and the difference is probably due to chance, not to the chipped edge on the die. Now the question becomes how big does $\chi^2$ have to be before it is likely that the difference is due to the chip in the die and not from random variation.

One way to answer this question is to look at chi squares generated from purely random events where the differences are known to be due only to random variation. In other words, toss a die *you know to be fair* sixty times and calculate the chi square for the outcomes. One chi square is not a large enough sample to use for comparison, however, so the experiment must be repeated enough times to have a fairly large distribution of chi squares. Again, tossing a die sixty times becomes very boring. An alternative is to use a calculator, a chart, or a simple computer algorithm to simulate random tosses of a die; read sixty numbers, record the number of times each face occurs, and calculate the chi square. If everyone in class does this, and all the results are combined in one stem-and-leaf plot, a distribution like the one in table 6.3 will result.

Table 6.3

| $\chi^2$ | Results |
|---|---|
| 0 | 244 |
| 1 | 0246666 |
| 2 | 00244446888 |
| 3 | 226 |
| 4 | 68 |
| 5 | 0266 |
| 6 | 0224468 |
| 7 | 004 |
| 8 | 246688 |
| 9 | 8 |
| 10 | 2 |
| 11 | |
| 12 | 02 |

8|2 = 8.2

There are 13 numbers $\geq$ 7.0.

How does the original chi square fit into the distribution? What is the probability that a chi square of 7 will occur by chance? The number of times a chi square greater than or equal to 7 occurred by chance is the total number of chi squares greater than or equal to 7 divided by the total number of chi squares that are calculated. In this example there are 13 out of 50 chi squares greater than or equal to 7, so $P(\chi^2 \geq 7) = .26$. This means that about one-fourth of the time a chi square of 7 or larger would occur by chance. This is not a very unusual probability, so it seems that for this experiment, 7 is really not a very large chi square. The difference is most likely due to chance and not to the chipped edge on the die.

When will a chi square be "large enough"? This depends on what definition of an unlikely event is used. If something that happens 1 out of 10 times is considered unlikely, then a chi square that occurred less than 10 percent of the time might be significant. For some cases, a chi square might have to occur less than 5 percent of the time to be interesting.

A "one tailed" 90 percent box plot (table 6.4) can be constructed from the chi square distribution for tossing the die by finding a value below which at least 90 percent of the experimental values fall. In the example, at least 90 percent of the fifty trials means the box would have to contain at least forty-five of the outcomes. Another way to think of this would be that five or fewer results would be outside the box. Because the fifth and sixth values in the distribution are both 8.8, only four values can be outside, which is why the words "at least" are used for describing the number of values inside the box. This means that at least 90 percent of the time a random chi square for a fair die should be between 0 and 8.8. Only 10 percent of the time would a chi square greater than or equal to 8.8 occur due to chance. The chi square of 7 for the experiment is less than 8.8 and thus is not "significant" at the 10 percent level. The conclusion is that the difference between the expected and observed results was probably due to random variation and not inherent in the die.

Table 6.4

| $\chi^2$ | Results |
|---|---|
| 0 | 244 |
| 1 | 0246666 |
| 2 | 00244446888 |
| 3 | 226 |
| 4 | 68 |
| 5 | 0266 |
| 6 | 0224468 |
| → 7 | 004 ← [Observed $\chi^2$ = 7] |
| 8 | 246688 |
| 9 | 8 |
| 10 | 2 |
| 11 | |
| 12 | 02 |

8|2 = 8.2

Note that chi square is a number that is unit free. It has no meaning unless analyzed in the context of the corresponding chi square distribution. A more detailed explanation of chi square, including the conditions under which it is a valid statistic, the impact of the number of variables, and table values, can be found in most introductory statistics texts. The following activity, "Which Cola Do You Prefer?" can be used in your classroom to furnish an interesting introduction to statistical inference using chi square.

*Try This:* ***Give the following project to a class. Ask twenty students to choose a color from red, blue, green, and yellow. Calculate the chi square and a chi square distribution for four choices. Decide if the students in the sample randomly selected a color and justify your conclusion.***

*Assessment Matters:* ***Be sure that your tests on this material are consistent with your approach to instruction. For example, on your tests expect students to discuss the meaning of certain chi square values not just compute them. If students use computer software in class and on homework, allow them access to similar tools on your tests.***

*Try This:* ***Have students use the computer program (p. 59) to investigate the following questions. How are the chi square distributions for three, four, and five outcomes alike and how are they different? What happens to the distribution as the sample size changes?***

## A TASTE TEST: WHICH COLA DO YOU PREFER?

**Teacher's Guide**

EXPERIMENT: Do students as a class have a definite preference for a certain brand of cola?

EQUIPMENT: Three liters of different colas, preferably in plastic bottles, and small paper cups

PROCEDURE: Remove the labels from each of the colas and mark one bottle X, another Y, and the third Z. Students should mark the side of their cups X, Y, Z. (If the number of students taking part in the experiment is divisible by 3, the computation will be simplified. If necessary, students can solicit tasters from another class.) To randomize the order of testing, groups of students should taste in various orders, such as X, Y, Z; X, Z, Y; Y, X, Z; and so on. Three pourers should pour a small amount of soda into the appropriate cup for each student. The students should not taste the soda until all the cups are full and should vote only for a single preference.

Tally the results of the class votes on the activity sheet. After all votes are in, calculate the chi square. The expected results would indicate there was no real preference, and the votes were random. For example, if there were twenty-four students tasting the colas, each cola would have eight votes.

To provide a standard to use for comparison, use a fair die to simulate the random tasting. Let faces 1 and 2 be brand X, faces 3 and 4 be brand Y, and faces 5 and 6 be brand Z. Roll the die once for each student who tasted the cola and record the face (letter). Complete the table for the random votes and calculate the chi square. Record each student's random chi square on the stem-and-leaf distribution. The table may need to be extended if students generate large chi squares. (In order to get a large enough sample of chi squares, some students may need to calculate several random chi squares.) Have students locate the chi square from the cola test on the distribution and calculate the probability that a chi square is greater than or equal to this number. Their conclusion should be based on their interpretation of this probability and what they perceive as being a "significant" chance.

Students may raise a variety of considerations. Do plastic bottles change the taste of the colas? Does the temperature of the soda influence the taste? Does the amount of soda tasted make a difference? These are valid concerns and might affect any real decision based on the results. Have students recommend which brand of cola the school soda machine should stock and why. As a final step, try to identify the brands of soda.

As an extension, identify several places where the chi square test might be used. Consider areas such as testing, jury selection, or science. A biology teacher might have some real applications from class experimentation.

The computer program on the following page can be used to calculate the chi square for *n* outcomes of an experiment.

## A BASIC Program to Simulate a Chi Square Distribution

```
LIST
 90  TEXT : NORMAL : HOME
100  PRINT "SIMULATION OF CHI SQUARED DISTRIBUTION"
105  PRINT : PRINT : PRINT : PRINT
110  PRINT "HOW MANY CHOICES ARE THERE";
120  INPUT C
125  PRINT : PRINT : PRINT
130  PRINT "WHAT IS THE SAMPLE SIZE";
140  INPUT N
150  EX = N / C
154  PRINT : PRINT : PRINT : PRINT
155  INPUT "ENTER NUMBER OF TRIAL SIMULATIONS==>";NS
160  DIM K(C),CHI(NS)
165  HOME
170  FOR Z = 1 TO NS
180  FOR Y = 1 TO C:K(Y) = 0: NEXT Y
190  FOR T = 1 TO N
200  Y = INT (RND (1) * C + 1)
210  K(Y) = K(Y) + 1
220  NEXT T
230  CHI(Z) = 0
240  FOR Y = 1 TO C
250  X = ((K(Y) - EX) ^ 2) / EX
260  CHI(Z) = CHI(Z) + X
270  NEXT Y
290  PRINT "TRIAL # ";Z;"  CHI SQUARE = ";CHI(Z)
300  IF Z / 2 = INT (Z / 2) THEN 320
310  INVERSE : GOTO 330
320  NORMAL
330  NEXT Z
335  NORMAL
340  PRINT "WHAT IS YOUR COMPUTED CHI SQUARE";
350  INPUT S
360  P = 0
370  FOR Z = 1 TO NS
380  IF CHI(Z) > = S THEN P = P + 1
390  NEXT Z
394  LET P = (P / NS) * 100
396  LET P = INT (10 * P + .5) / 10
400  HOME
410  VTAB (10) : PRINT "THE PROBABILITY OF YOUR RESULTS BEING"
420  PRINT : PRINT "DUE TO CHANCE IS ";P;"%"
500  REM  CREATE NIGHTINGALE DATA FORMAT FOR STEM AND LEAF PLOT
510  PRINT : INPUT "HIT 'RETURN' TO SEE STEM AND LEAF PLOT" ;ZZ$
520  PRINT : PRINT "                WORKING..."
525  REM  SORT THE DATA
530  FOR I = 1 TO NS - 1
540  FOR J = 1 TO NS - 1
545  IF CHI(J) < = CHI(J + 1) THEN 560
550  LET D = CHI(J) : LET CHI(J) = CHI(J + 1) : LET CHI(J + 1) = D
560  NEXT J
570  NEXT I
600  REM  CREATE AND OPEN THE FILE
604  REM  LET NR=NS TO AVOID CONFUSION WITH STEM AND LEAF PROGRAM
605  LET NR = NS
610  LET D$ = CHR$ (4)
620  PRINT D$;"OPEN SIM.DATA"
630  PRINT D$;"WRITE SIM.DATA"
640  PRINT 2
650  PRINT "SIMULATED VALUES OF"
660  PRINT "CHI-SQUARE"
670  PRINT "DATE" : PRINT "880110"
680  PRINT "NUMBER OF RECORDS": PRINT NR
690  PRINT "NUMBER OF VARIABLES": PRINT 1
700  PRINT " ": PRINT " "
710  FOR I = 1 TO NS
720  PRINT CHI(I)
730  NEXT I
740  PRINT D$;"CLOSE SIM.DATA"
800  REM CREATE THE NULL FILE
810  LET ND = 1
820  LET N$ = "BLANK"
830  PRINT D$;"OPEN NULL.FILE"
840  PRINT D$;"WRITE NULL.FILE"
850  PRINT ND: PRINT N$: PRINT N$
860  PRINT D$;"CLOSE NULL.FILE"
870  PRINT D$;"RUN STEM AND LEAF"
999  END
```

***ACTIVITY 18***

***A TASTE TEST: WHICH COLA DO YOU PREFER?***

SHEET 1

1. After tasting the three colas, record the class preferences in the chart.

Tally of votes

Brand X ____________________

Brand Y ____________________

Brand Z ____________________

2. Calculate the chi square for the class results.

Chi Square from Colas

| Colas | Observed | Expected | (O – E) | $(O - E)^2$ | $(O - E)^2/E$ |
|---|---|---|---|---|---|
| X | | | | | |
| Y | | | | | |
| Z | | | | | |

$\chi^2$ = Sum $(O - E)^2/E$ = __________

3. Use a die you know to be fair or use a random die chart to find at least one chi square.

Chi Square Calculated from Fair Die

| Faces | Observed | Expected | (O – E) | $(O - E)^2$ | $(O - E)^2/E$ |
|---|---|---|---|---|---|
| 1, 2 (X) | | | | | |
| 3, 4 (Y) | | | | | |
| 5, 6 (Z) | | | | | |

$\chi^2$ = __________

4. Make a distribution of the random chi square calculated by each member of the class.

Chi Square Distribution

| |
|---|
| 0 |
| 1 |
| 2 |
| 3 |
| 4 |
| 5 |
| 6 |
| 7 |
| 8 |
| 9 |
| 10 |
| 11 |
| 12 |
| 13 |
| 14 |

5. Construct a one-tailed 90% box plot of the distribution.

0
1
2
3
4
5
6
7
8
9
10
11
12
13
14

6. The estimated probability that chi square is ≥_____ = $\dfrac{\textit{the number of chi squares} \geq \_\_\_\_\_}{\textit{the total number of trials}}$ = _____.

7. Is there persuasive evidence of a preference for a certain soda among those tested? Why or why not?

# CHAPTER 7
# STUDENT PROJECTS

"Providing opportunities for discussions about issues, people, and the cultural implications of mathematics reinforces student understanding of the connections between mathematics and our society" (NCTM1989, p. 140). Nowhere in the secondary school curriculum is this statement more relevant than in teaching statistics. Statistics is a tool that is used daily in the world that surrounds students. A vital part of learning statistics is student involvement in the collection, organization, and evaluation of data that come from the student's environment. Equally important is the student's development of skills in communicating the results of analyzing data. Statistical techniques come alive when students use them with their own data. Students learn to reason and make decisions about issues, large or small, by designing and carrying out projects. Project work, however, is often difficult for both teachers and students. For many students, doing a statistics project is their first encounter with applying mathematics to a real-world situation. Students resist having to write in a mathematics class. They think writing is something they do only down the hall in English class. Teachers frequently have to coax students to meet project deadlines and work with them on how best to collect data for their area of interest.

This chapter presents ideas for both short and extended student projects. Ideas and resources for such projects are included along with samples of student work and possible evaluation schemes.

## MINIPROJECTS

**ASSIGNMENT A: Write a personal reaction to an article about the use of statistics in research (fig. 7.1).**

# Researcher to use death camp data

*Use of Nazi studies draws sharp criticism*

Associated Press

Minneapolis—A scientist is planning to use Nazi studies of concentration camp prisoners deliberately frozen to death to further his research about hypothermia.

But some scientists and Jewish leaders have sharply criticized his intention to use data obtained during the Holocaust.

Dr. Robert Pozos, director of the hypothermia research laboratory at the University of Minnesota, Duluth, says he plans to analyze and republish "The Treatment of Shock from Prolonged Exposure to Cold," a study by doctors at Dachau. It includes observation and physiological measurement of people placed in vats of freezing liquid, often to the point of death, according to those familiar with the study.

"We should under no circumstances use the information. It was gained in an immoral way," said Dr. Daniel Callahan, director of the Hastings Center, a medical ethics think tank in Briarcliff Manor, N.Y.

"I think it goes to legitimizing the evil done. I think the findings are tainted by the horror and misery," said Abraham H. Foxman, national director of the Anti-Defamation League.

Because mammals differ widely in their physiological response to cold, hypothermia research is uniquely dependent on human test subjects, says Pozos, a specialist in the field for 12 years.

Several medical ethicists contend study of the Nazi research could save lives and, if published with a condemnation of the methods, call attention to the plight of the Jews, Poles and Gypsies killed in the experiments.

Because the Egyptians used slave labor, "does that mean we should never gaze at the pyramids?" asked Dr. Thomas Murray, director of the Center for Biomedical Ethics at Case Western Reserve University in Cleveland.

**Fig. 7.1**

♦ ♦ ♦ ♦ ♦ ♦ ♦ ♦

STUDENT RESPONSE: After reading the article concerning the use of the Nazi Death Camp Data, I feel the results should be used for further research of hypothermia. I feel this way despite the fact that the data was collected in an inhumane manner. As I see it, the act has been committed and it should be used to help others in the case of hypothermia.

GRADE AND RATIONALE: Just fulfilled the requirements as personal statement with a brief rationale. Perhaps more on why you think it will help others; is this the only way to collect the data? Would the data be biased in any way? "B–"

---

***ASSIGNMENT B: Write a summary of an article from a magazine, journal, textbook, or other source that shows a practical application of statistics or how statistics is used in society. This is to be typed or written in ink and double spaced. Also, include a cover page and complete bibliography of the source article. Indicate in your essay the date of publication and discuss what effect this may have on the information.***

STUDENT RESPONSE: ELECTION POLLING

As long as the United States has existed, the people (to varying degrees) have had a vested interest and a general curiosity about who won the most recent election. As technology progressed in proportion to people's expectations thereof, the populace demanded results faster and faster—so fast that now the television news networks have projections the minutes the polls close around the country. In fact, newspapers and television networks were highly criticized during the 1988 Presidential election for projecting winners weeks and months in advance, always updating at the slightest "change" in the public opinion polls.

Polls in one form or other have existed as long as man has walked upright, but the election night poll really did not gain prominence until the early and middle 20th century, notably when the pollsters incorrectly predicted the defeat of F.D.R. and later of Truman (both proved incorrect). Television began to report elections in 1952 (Link, 179) and primaries were covered also by 1964. 1964 marked a watershed year for television networks and their pollsters—it was said that more pollsters were in New Hampshire during the primary than their were voters, or that it would have been cheaper for the networks to bring the voters to New York City to cast their ballots (Link, 180). Afterwards, the three networks and two major news services founded a clearing-house organization called the News Election Service (NES) whose mission was to release uniform data to all news and others services at the same time with same data. NES provides such information for national (Presidential, House, and Senate) and state elections.

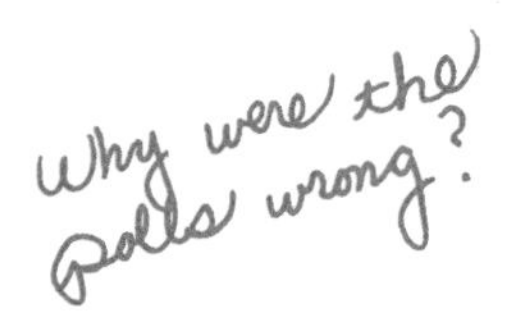


NES utilizes computers and key precincts to calculate projections and percentages. Each state, county, and national election has its key precincts, a key precinct being a precinct whose demographic factors and voting record is well-documented and fairly "predictable." Election reports compare key precinct polling-percentages with county, state, and national and calculate predictions on the basis of that and previous known political breakdowns. NES can take these precedents in the form of percentages, such as knowing that usually 60% of New York City votes Democratic, and multiply them by the early election returns for the state and its pecentages, and arrive at a total state-wide prediction. For Presidential elections, electoral

clear

◆ ◆ ◆ ◆ ◆ ◆ ◆ ◆

votes play a key role in the mathematics of the project. Computers are used to create complex models using such precedental and actual percentages, which then let NES release data and predictions.

Also done on election night, is attitude and issue polling—questions about an individual's feelings about current important issues. These results are analyzed to determine how particular demographic groups vote, such as age or race groups. Altogether, this election night polling of votes and issues is getting more accurate as it is getting more widely used.

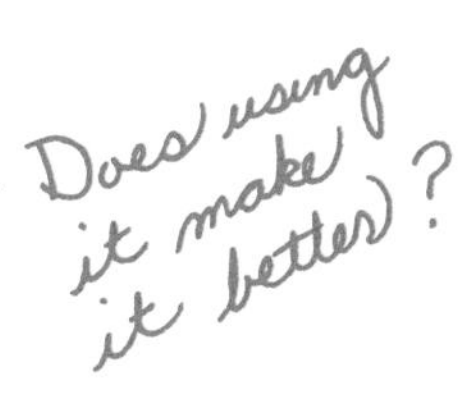


Bibliography

Link, Richard F. "Election Night on Television." Statistics: A Guide to the Unknown, 2nd Ed. San Francisco: Holden-Day, 1978. pp. 178-186.

GRADE AND RATIONALE: *The report presents a nice view of the history of polling but does not touch much on the statistics involved. Why were the polls wrong in predicting the defeat of Truman and FDR? What did they do to prevent this from occurring again? Much has changed since the article was written in 1978—do you think there has been any change in the polling process? Proofread your work a bit more carefully. Short sentences that make a point are better than long, involved ones—hard to follow the logic in some. "B"*

---

***ASSIGNMENT C: Collect an example from a newspaper or magazine in which a graph has been presented in a potentially deceptive manner. Identify the source from which the graph was taken. Explain briefly the ways in which the graph might have been deceptively presented and then suggest ways the data might be presented more fairly or in a less distorted fashion. An original or photocopy of the graph must be included with the project.***

STUDENT RESPONSE: This graph is distorted in two ways. First of all, the Best Car Rental Franchisees are shown in white, a color which makes objects seem larger. The other Franchisees are shown in a dark graph color which blends with the black background making the human figures appear smaller. Secondly, the sizes of the human figures are increased by length as well as width increasing the figures unproportionally in area.

GRADE AND RATIONALE: *Fulfilled assignment requirement well and corrected graph is a nice way to get the message across using the same technique. "A"*

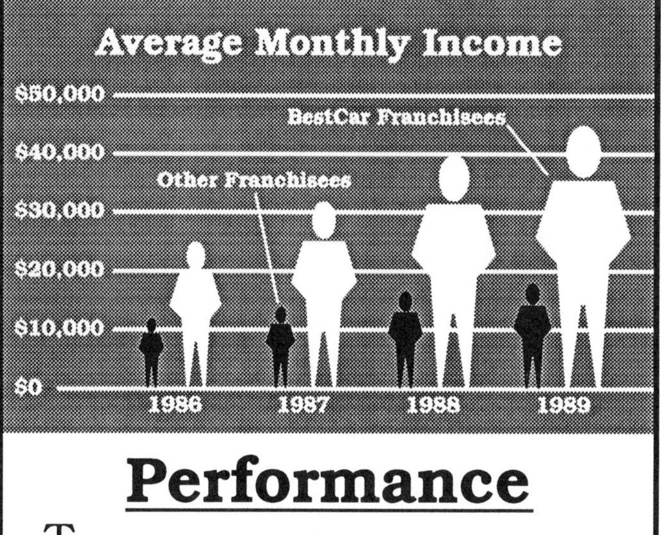


**Performance**

There is an outstanding company in the car rental industry!

Franchise owner incomes for this company, committed to excellence, are 250% of the industry average and at least 70% higher than their closest competitor. This company is not only #1 in sales but outranks everyone in service and customer support as well. The company attracts the best business people and provides them with a proven opportunity to achieve both professional and personal goals. This company, which will not settle for second best, is looking for more franchise owners.

For information on franchise opportunities, write PO Box 115. Minimum cash: $60,000 on total investment of $240,000.

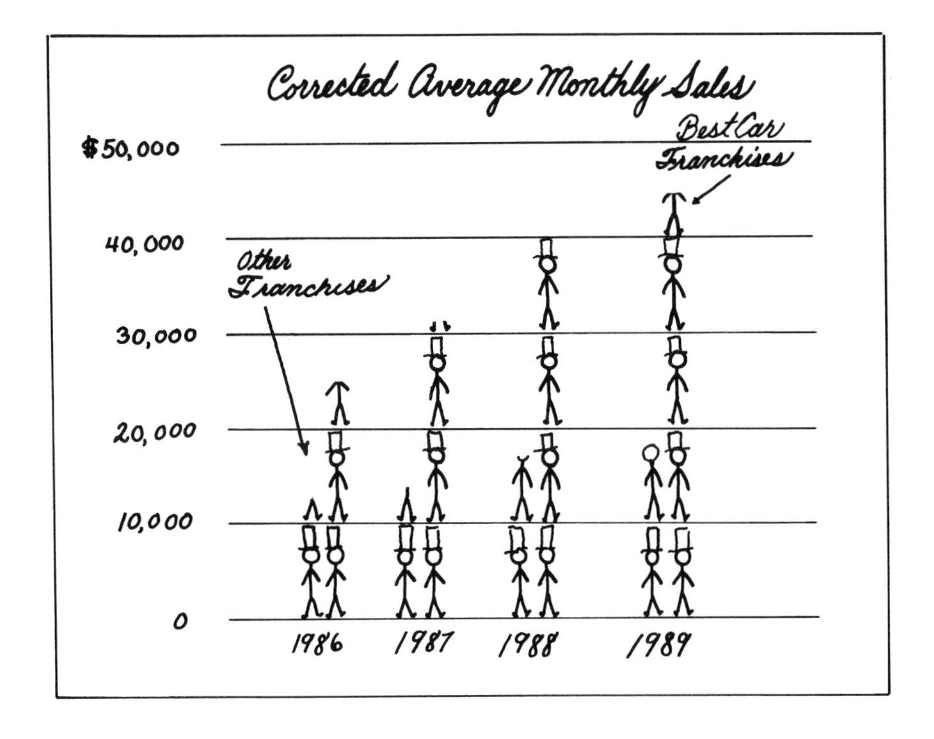


***ASSIGNMENT D: Collect two data sets. Use a back-to-back stem-and-leaf plot or multiple box plots to compare them. Write a short description of your findings.***

STUDENT RESPONSE:

AGES OF MOTHERS AND FATHERS OF
THE 1990 BROOKSTONE SENIOR CLASS.

The majority of the mothers and fathers of the 1990 senior class are in their forties. The youngest mother is 39, while the youngest father is 38 years old. The oldest father is 63, while the oldest mother is 57 years old. The mean, or average, of the mothers ages is 44.6 years. The mean of the fathers ages is 46.8 years. The mean of both the fathers and mothers ages is 47.7 years. The median father age is 46. The median mother age is 43.5. Both plots have mound shape distributions. There are 50 mothers, but only 47 fathers due to the death of 3 fathers.

| Mothers | 3 | Fathers |
|---:|:---:|:---|
| | . | |
| | . | |
| | . | |
| 9 | . | 8 |
| 111111000 | 4 | 000111111 |
| 333333322222222 | . | 22223333 |
| 5555544444 | . | 44555 |
| 776666 | . | 6666777 |
| 888 | . | 8899 |
| 0 | 5 | 0111 |
| 22 | . | 223 |
| 55 | . | 5 |
| 7 | . | 7 |
| | . | 9 |
| | 6 | |
| | . | 223 |

5 | . 2 = 52 years

♦ ♦ ♦ ♦ ♦ ♦ ♦ ♦

GRADE AND RATIONALE: The following key will be used to grade each assignment

| | |
|---|---|
| 10 | 5 points for each data set |
| 10 | 10 points for the graph |
| 0 | 2 points for the data source |
| 3 | 6 points description |
| 3 | 6 points for the comparison |
| 3 | 6 points for rationale |

*How was the data collected and what was the source? The description gives too many statistics to keep track of. Select one measure of central tendency and use it to describe the age spread between mothers and fathers. The plots would more accurately be described as skewed to the low end with a long tail.*

Total number of points earned: 29 ; Total number of points possible: 40

(Note that the point allocation does not have to become a percent grade for passing/failing. 29/40 points indicates the response is less than 3/4 of what was anticipated. This can be assigned any grade or factored into a grading system in a variety of ways.)

SECOND STUDENT RESPONSE: Using data from the World Almanac and Book of Facts, we examined the ages of men and women who won Oscars between 1928 and 1987. From this we reached the following conclusions:

- The median age for actresses is 33 years.
- The median age for actors is 41 1/1 years.
- We considered that perhaps actresses are judged partially on physical appearance. If younger actresses are considered more attractive, this might contribute to the lower median age.
- We also feel that competition for men is stiffer than that for women because there are more male leading roles available. To be a winning actor, more experience is necessary thus raising the median age.

| Actor | | Actresses |
|---:|:---:|:---|
| | 2 | 1244444 |
| | . | 566666677788999 |
| 44332210 | 3 | 0000112333444444 |
| 99888888775555 | . | 55677888 |
| 4433332211110000 | 4 | 0111112 |
| 99988877665 | . | 589 |
| 321 | 5 | |
| 6665 | . | |
| 210 | 6 | 0112 |
| | . | |
| | 7 | 4 |
| 6 | . | |

Age when each actor or actress won Academy Award (1928-1987)

Scale 3|2 means 32 years

Academy Award data was updated from World Almanac 1988 and Book of Facts published by Pharos Books New York, NY 10166.

♦ ♦ ♦ ♦ ♦ ♦ ♦ ♦

GRADE AND RATIONALE: The following key will be used to grade each assignment.

<u>10</u> 5 points for each data set

<u>10</u> 10 points for the graph

<u>2</u> 2 points for the data source

<u>3</u> 6 points description

<u>3</u> 6 points for the comparison

<u>6</u> 6 points for rationale

*To compare the data, include a listing and summary of the IQR and extremes as well as measures of the center. Do not just list statistics without indicating what they represent in words. Use these statistics to support your conjectures.*

Total number of points earned: <u>34</u> ; Total number of points possible: <u>40</u>

AN ALTERNATIVE STUDENT RESPONSE: In general actors win an Oscar when they are older than actresses who win Oscars. The males are usually between the ages of 35 to 45, and half of all the actors winning Oscars were older than 41. Most of the actresses, however, won their Oscar when they were younger, usually from 25 to 35 years old. The median age for females to win an Oscar was only 33, 8 years younger than the median age for actors. Henry Fonda was an outlier because he won an Oscar when he was 76, much older than any of the others. There were 5 actresses who won after the age of 60; the oldest one of these was Kathryn Hepburn who won with Henry Fonda for *On Golden Pond*. Perhaps actresses are judged partially on physical appearance. If younger actresses are considered more attractive, this might contribute to the lower median age. We also feel that competition for men is stiffer than that for women because there are more male leading roles available. To be a winning actor, more experience is necessary thus raising the median age. Actresses might be judged on the basis of their appearance, and that would contribute to winning the Oscar earlier.

Total number of points earned: <u>40</u> ; Total number of points possible: <u>40</u>

When students have had experience collecting, organizing, and analyzing data, they are ready to work on short projects. These projects can be assigned midway through a unit and collected several weeks after the unit to give students time to assimilate the concepts they covered. The project guidelines sheet (p. 67) can be handed out as a guideline for students when the short projects are assigned.

A 60-point scale such as that below can be used to grade projects.

5 Format
5 Problem pertinent, well defined
10 Collection of data
5 Discussion of possible bias
10 Numerical statistics appropriate and explained
10 Graphical displays appropriate and explained
15 Description of conclusion including use of statistics to justify results.
60 Total points possible

Long-range projects are an appropriate way to determine whether students can bring together all the ideas they have been studying. For this reason, these projects are particularly timely when due at the end of the semester. Pages 68–70 provide a sample student abstract and guidelines and suggestions that can be given to students when a long-range project is assigned.

## SHORT-TERM STATISTICS PROJECT GUIDELINES

### Assignment

Through observation, research, or experimentation, compile a list of sample data. Obtain at least 40 values. Try to select data from an interesting or meaningful population.

a. Describe the question investigated and the nature of the data. That is, what do the values represent?

b. Describe the method used in collecting the data.

c. Explain possible reasons why the data might not be representative of the population. That is, what are some possible sources of bias or error?

d. Calculate appropriate statistics from the following: sample size, minimum, maximum, mean, midrange, range, standard deviation, variance, and the quartiles $Q_1$ and $Q_3$ and discuss their relation to the data.

e. Construct a frequency table, stem-and-leaf plot, box-and-whiskers plot, or histogram when appropriate and explain what each tells you about your data.

f. Write, in paragraph form, any conclusions or inferences that can be made from an analysis of your data.

### Format for Statistics Short Projects

1. Each report should begin with a cover sheet that includes the title of your project, your name, and the date.
2. The report should be typed or written neatly in ink. (Double space both written and typed reports.) Charts, data sheets, and graphs should be neatly constructed and written in ink. They do not have to be typed. Graphs and charts should have a title and properly labeled axes.
3. Include computer runs with the project when applicable.
4. List all references used (title, author, publisher, and copyright date) for the project.
5. All mathematical formulas must be given when computing test statistics. The substitution step should be shown (when possible) before numerical answers are given. When using variables, be sure to state what each variable represents.
6. All explanations must be in paragraph form.
7. When organizing your project, place data and graphs immediately preceding the bibliography.

### Project Checklist

1. Title due ____________________
2. Rough draft due ____________________
3. Final draft due ____________________

## LONG-TERM STATISTICS PROJECT GUIDELINES

1

There are some basic steps one needs to consider in order to produce a successful statistics project. A guideline of what you should include in a project is listed below.

I. You need a question or problem. Statistics is a tool to help you answer the question.

II. Define the problem or question in clear, specific terms.

III. Develop hypotheses.

IV. Find out as much as you can about the question.

A. Has someone already done work on this question?

B. Is the question one for which there is an answer?

C. Collect research about your topic. (Be sure to include this information in your Works Cited.)

V. Design the study and develop techniques and measuring instruments that will provide objective data pertinent to the hypotheses. (Remember that all surveys must be field tested before use.)

VI. Collect the data.

VII. Analyze the data.

VIII. Interpret the results and draw conclusions relative to the hypotheses based on the data.

A. Use charts, tables, histograms, box plots, stem-and-leaf plots, correlation lines, comparison of means, proportions—whatever is appropriate to display the data.

B. Remember to identify the sample size, distribution, and confidence level so that you use an appropriate test statistic.

IX. Write the results. Write and rewrite and....

### SAMPLE STUDENT ABSTRACT FOR A LONG-RANGE PROJECT

With the number of divorces constantly increasing, and the number of children involved in these divorces increasing also, there have been many speculations and studies on what effect divorce has and will have in the development of these children of divorce. Cumulative findings sponsored by the Foundation for Child Development indicate that divorce does serve to hinder the mental development of children whose parents divorce when they are young. (Diamond, 1985) Young children tend not to see divorce as a relief from family stress, and therefore their emotional distress often effects their school performance. (Murphy, 1983) Their efforts to do well fall off, their grades drop, their school attendance declines, and their behavior often gets worse. (Diamond, 1985)

In an effort to see if this claim could be substantiated by data, I administered a survey of children in four elementary schools, asking if their parents were married or divorced, their grade, number of absences, and an evaluation of their overall behavior. I believe these factors will indicate whether or not divorce has an effect on a child's performance at school.

## *FORMAT AND STRUCTURE FOR LONG-RANGE STATISTICS PROJECTS*



Format for Written Project

1. The project must be typed and double spaced.

2. All charts, data sheets, and graphs should be neatly constructed and prepared in ink or on a computer. They do not have to be typed. Graphs and charts should have a title and properly labeled axes.

3. All mathematical formulas must be given when computing test statistics. The replacement values and substitution (when possible) should be shown before numerical answers are given. Be sure to state what each variable represents.

4. All pages should be numbered.

5. Order of pages:

   a. Cover Sheet—The first page of the project must be a cover sheet that includes your name, date, teacher's name, and the title of the project.

   b. Table of Contents

   c. Hypothesis, Abstract, or Explanation of Project—A statement on why your project is important or relevant as well as your research goes here. You should also include your null and alternative hypotheses in this section.

   d. Explanation of Data—This should include how the data were collected or what research was done to obtain the data. Remember to document all items that need to be cited.

   e. Analysis of Data—This is where you include all statistical tests. Your graphs and charts go here.

   f. Problems in the Project—This might include some of the difficulties that occurred when collecting or analyzing data, as well as what you might want to change in a follow-up to the project. Limitations of your inferences should also be discussed.

   g. Conclusion—This is where you interpret your findings (what you can conclude or not conclude from your research).

   h. Discussion—Suggestions for further work in the area of your topic as well as recommendations should be mentioned here.

   i. Works Cited—Use a standard format for your paper.

   j. Appendix—This is where you include your data, sample surveys, cover letters, and listing of computer programs.

6. Final Presentation—The completed project, as well as an oral presentation, will be due two weeks before finals begin. An exact date will be announced in class.

7. It is a good idea whenever submitting a major project to keep a photocopy for your own files.

## EVALUATION

The statistics project will be evaluated on the following:

1. Creativity and originality
2. Organization and neatness (Does the report adhere to format guidelines?)
3. Clarity (vocabulary, English structure, data, charts)
4. Date of completion (All work must be submitted on or before the due date. If you are absent, your project must still be in school on the due date. There will be a one-letter-grade penalty for every day the project is late.)
5. Validity of conclusions (Was the objective accomplished?)

NO PROJECT COMPLETED AND TURNED IN ON TIME WILL RECEIVE A GRADE BELOW C.

Time line for project:

1. Title and hypothesis due week 9 of the semester
2. Questionnaires due for field testing week 11 of the semester
3. Data due week 13 of the semester
4. Rough draft due week 15 of the semester
5. Final draft due week 16 of the semester

Good sources of data:

1. *Wall Street Journal*
2. *Consumer Reports*
3. Almanacs
4. *Statistical Abstract of the U.S.*

## SUGGESTED IDEAS FOR PROJECTS



1. Do more students smoke in schools with smoking areas than in schools without smoking areas?
2. Do males or females have higher math grades in high school? (This can also be expanded to science grades.)
3. Do boys have higher math SAT scores than girls?
4. Which students eat in the cafeteria and which do not? Analyze the reasons from a survey.
5. Make predictions.
   a. Estimate the number of students in your school who ....
   b. Estimate the college grade point average from high school grades.
6. Do ninth graders study more than twelfth graders; are their grades better than twelfth graders'?
7. Is there a correlation between grade point average and
   a. amount of time spent watching TV?
   b. number of children in a family?
   c. number of hours spent studying?
   d. whether you work after school or not?
8. Do students in your school score higher on standardized tests than the national average?
9. Are insurance rates unfairly higher for males under twenty-five?
10. Does eating a good breakfast have any effect on school performance?
11. Do students who play a musical instrument have higher grades than the general school population?
12. Do more people prefer Coke over Pepsi (or vice versa)?
13. Is there a correlation between attendance in class and period of the school day?
14. Is there a correlation between grade point averages of students and participation in extracurricular activities?
15. Does smoking affect absenteeism in high school students?
16. Are parents more protective of their daughters than of their sons?
17. Is there a correlation between the grade in a subject and the period the subject is taken?
18. Are people who remember their dreams more creative than those who do not?
19. AIDS: How informed are students?
20. Is the Dow-Jones average a good indicator of the economy?
21. Do the reaction times for males and females differ significantly?
22. Is there a correlation between hand guns sold per year and homicides per year?
23. What is the correlation between blood pressure and age?
24. What is the correlation between ACT scores and SAT scores?
25. Overall, which grocery store has the least expensive prices?
26. How are average class sizes determined? (See *Mathematic Magazine*, May 1982.)
27. How are minorities affected by the U.S. population census undercount?
28. Does race affect the choice between the use of nursing homes as opposed to in-home care of the elderly?
29. Can higher SAT scores be achieved by increasing study time or by taking a review course?
30. What is the relationship between income and SAT scores?
31. How does a student's grade from a teacher affect the student's opinion of that teacher?
32. Does the race of a student affect the way he or she spends discretionary money?
33. Do stock prices vary with the day of the week?
34. Construct a profile of a person who attends your school or lives in your community or state.
35. Compare the profiles of two sets of towns or groups of people.

## CHAPTER 8
## ASSESSING STATISTICAL UNDERSTANDING

Assessment is an essential part of the instructional activities and "fundamental to the process of making the *Standards* a reality" (NCTM 1989, p. 190). Usually tests are seen as something separate from instruction; passing a test is often seen as the goal of instruction. Tests, however, should be an integral part of mathematics instruction; teachers should not merely count correct answers to assign grades. Because statistics, in particular, is not a hierarchy of skills and algorithms, evaluating statistical understanding must reflect a different perspective, one that looks at many different facets of student progress. Other essential principles in assessing statistical understanding follow:

- Tests should allow students to show what they know (positive testing).
- Tests should be integral to and reflect all goals of the curriculum (not only specific and isolated skills).
- Tests should be given using a variety of techniques, including written, oral, and construction modes (examples follow).

A series of examples taken from experiments in the Netherlands and the United States and carried out by the Research Group on Mathematics Education (OW & OC) of Utrecht University illustrate certain points. Similar examples occur in the suggested student activities of chapter 2, "An Introduction to Data." The American experiences are the result of a joint project between OW & OC and the National Center for Research in Mathematical Sciences Education (NCRMSE) located at the University of Wisconsin—Madison.

### COLLECTING DATA

*Example 1*

A magazine for health foods and organic healing wants to establish that large doses of vitamins will improve health. They ask readers who have regularly taken vitamins in large doses to write in, describing their experiences. Of the 2754 readers who reply, 93 percent report some benefit from taking vitamins.

- Is the sample proportion of 93 percent probably higher than, lower than, or about the same as the percent of all adults who would perceive some benefit from large vitamin intake? Why? (Steen 1988)

*Example 2*

A researcher wants to find out how many Americans intend to spend their holidays in the United States this year. To avoid bias, she does not want only volunteers to fill out the questionnaire. The researcher chooses the following strategy: she visits 27 travel agencies in big cities (because of the high concentration of inhabitants) and will interview every seventh visitor. The results of her research were published and entitled "Record Number of Americans to Foreign Destinations."

- Do you think that was a good way to set up the research? What kind of improvements do you suggest? (de Lange and van der Kooij 1989)

It will be clear that these two examples will lead to interesting discussion in the classroom. What is uncommon for many teachers, and students as well, is the fact that the answer (or more precisely an answer) is not a number and that there are many proper answers. This leads, of course, to a discussion about how to establish a "proper" answer and how to grade it. It is clear that problems like these two form an essential part of any statistics curriculum.

These two problems also focus on a more general goal of mathematics education but even more so for statistics education: a critical attitude. One of the main aims of both should be to encourage a critical attitude toward the results and outcomes in which mathematics and statistics have been used, especially in the media (de Lange and Kindt 1985).

***CRITICAL ATTITUDE***

*Example 3*

In a certain country the defense budget was 30 million dollars for 1980. The total budget for that year was 500 million dollars. The following year the defense budget was 35 million, whereas the total budget was 605 million dollars. Inflation during the period between the two budgets was 10 percent.

- You are invited to hold a presentation for a pacifist society. You want to explain that the defense budget has decreased this year. Explain how to do this.
- You are invited to hold a presentation at a military academy. You want to explain that the defense budget has increased this year. Explain how to do this.

This exercise gave rise to an interesting discussion between teachers. Some teachers said that it should indeed be our aim to develop a critical attitude but that we should not teach our students how to manipulate. Others stated that you can learn to identify manipulation by others only if you can manipulate data yourself.

*Example 4*

In 1966, 223 people applied for 15 jobs. They all had to take a test. The histogram in figure 8.1 shows the resulting scores of all 223 people.

- Try to explain the large gap in the graph. (Freedman, Pisani, and Purves 1980)

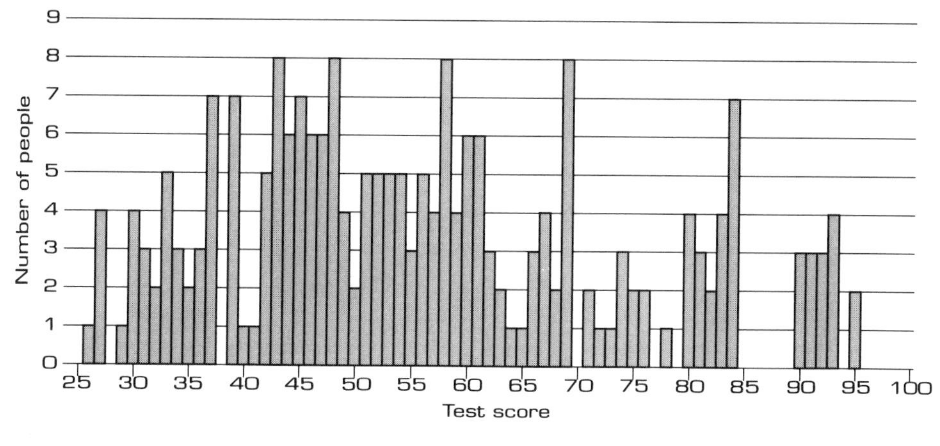


**Fig. 8.1**

This was certainly not a very easy problem for both teachers and students alike. Only after focusing on the relation between the number of available jobs and the number of scores on the right side of the gap does the problem become clear.

*Example 5*

The results of two classes on a mathematics test are presented in a stem-and-leaf display shown in figure 8.2.

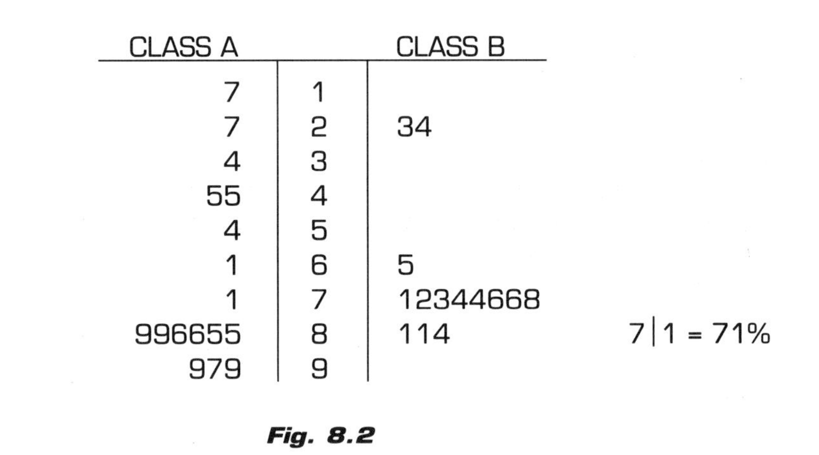

| CLASS A | | CLASS B | |
|---|---|---|---|
| 7 | 1 | | |
| 7 | 2 | 34 | |
| 4 | 3 | | |
| 55 | 4 | | |
| 4 | 5 | | |
| 1 | 6 | 5 | |
| 1 | 7 | 12344668 | |
| 996655 | 8 | 114 | 7 \| 1 = 71% |
| 979 | 9 | | |

***Fig. 8.2***

Is this table sufficient to judge which class performed better? (de Lange and Verhage 1989)

This example shows that a very simple problem may not have a very simple solution. The teacher has to listen carefully to the discussion in class and has to accept different solutions. At first glance many students—and teachers as well—tend to find class A better than class B because of the many scores in the 80s. However, some students argue that class B is definitely better because only two students did really poorly. Again others argue that you cannot say anything because the average of both classes seems to be the same. However, the median is definitely better for class..., and how about the mode? So it boils down to the question, What do you mean by "better"?

### TESTS

As one can expect by now, it will be clear that assessing statistical understanding is about as complicated, exciting, confusing, and rewarding as teaching and learning the subject. Of course, attention should be paid to the fundamental ideas, by assessments like these:

- Draw a histogram, line graph, stem-and-leaf plot, and so on.
- Describe a data set by a summary statistic, such as the mean or median.

But at least equally important are evaluative tasks, such as these:

- What graphical representation is most appropriate for this set of data?
- How can I design a proper graph?
- Is this conclusion really based on these data?

♦ ♦ ♦ ♦ ♦ ♦ ♦ ♦

The latter activities are "productive"; the students are expected to organize and describe the data.

These activities should also be present in our tests. So besides simple restricted, timed, written tests, other means for assessment should be used. Timed written tests offer some opportunities for testing higher-order thinking skills; one just has to look at the examples. But other ways are needed also.

#### TWO-STAGE TESTS

The first stage is carried out like a traditional timed, written test. The students are expected to answer as many questions as possible within a fixed time. After having been graded by the teacher, the tests are handed back to the students and the scores are disclosed. Now the second stage takes place: given information about their mistakes, students repeat their work at home on their own, without restrictions.

#### THE ESSAY

In our experiments with teaching statistics, we found that a certain kind of essay was very successful. Students were given an article from a newspaper with much information in tables and in numbers and were asked to rewrite the article making use of graphical representations. The article dealt with the problem of overpopulation in the Republic of Indonesia. One of the graphics designed by students (fig. 8.3) shows clearly the large differences in the population over the islands.

The left bar on each island shows the area of the island as a percentage of the total area. The right bar shows the population of the island as a percentage of the total population. It is clear that Java is overpopulated but that some of the other islands have lots of room.

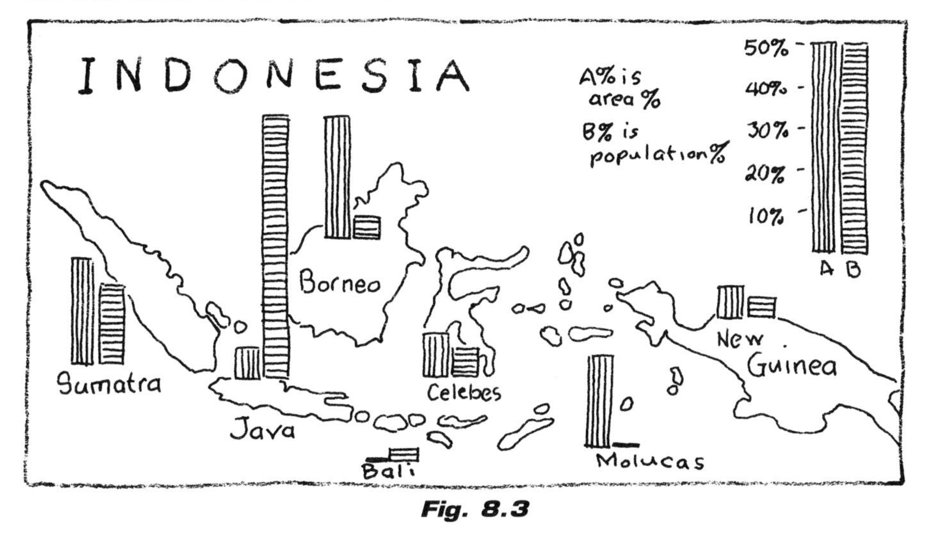


**Fig. 8.3**

#### THE STUDENT-MADE TEST

One of the more promising new ideas seems to be the following form for a test:

*Example 6*

This task is a very simple one (taken from an experiment carried out by OW & OC and NCRMSE at Whitnall High School in Greenfield, Wisconsin,

♦ ♦ ♦ ♦ ♦ ♦ ♦ ♦

in 1989). You have now completed the first two chapters of the book and have taken a relatively ordinary test. This test is different:

- Design a test for your fellow students that covers the entire booklet.
- You can work on your preparations from now on: look at magazines, papers, books, and so on, for the data, charts, and graphs that you want to see.

Keep in mind:

- The test should be taken in one hour.
- You should know all the answers.

The results of such a test are surprising. The test forces the students to reflect on their own learning processes and gives the teacher an enormous amount of feedback on his or her teaching activities.

#### CONCLUSION

There are a variety of ways student assessment in statistics can be carried out. There are formal tests, take-home tasks, and modified tests as indicated in this chapter. Student projects, of both short and long range, as discussed in chapter 7, provide another method of measuring student understanding. Essentially, assessment should measure what the students know and can do, which means that assessment instruments must provide students with the opportunity to demonstrate what they know and can do.

## *SOLUTIONS AND COMMENTS FOR ACTIVITIES*

### *Activity 2*

See Woodward and Ridenhour (1982) for another description of this activity and Gnanadesikan, Shaeffer, and Swift (1987) for information on simulation.

1–3. Solutions will vary.

4. Breaking the spaghetti is usually not random. This makes the first estimate quite different from that obtained by simulation.
5. The pieces will form a line segment when the sum of two of the lengths equals the third length.
6. The pieces will form a triangle when the sum of any two lengths is greater than the third.
7. The sum must be less than 100 in order to have a positive number for the third length.
9. a. $y > 50 - x$ b. $y < 50$ d. $x < 50$
11. Area equals 1250 square units.
12. The probability is 1250/5000, or 1/4.

### *Activity 3*

The following solutions are samples of answers that might occur.

1. The population of Alaska predicted for 1990 will range between 333 300 and 434 502 persons.
2. A negative percent would indicate the loss of $n$ persons per 100 people.
3. The probability of being incorrect is very high when predicting a specific value. However, the chance of being correct is increased considerably by predicting a result within a set of parameters.
4. Answers will vary with home state, that is, in Alabama the projected change in population is a growth between 5.0% and 9.9%.
5. There are 17 states that have a projected population growth between 5.0% and 9.9% in the 1990 census.
6. The states that are projected to have the largest percent gain in population are Alaska, Arizona, California, Colorado, Delaware, District of Columbia, Florida, Georgia, Hawaii, Maryland, Nevada, New Hampshire, New Mexico, North Carolina, Texas, Utah, Virginia, and Washington.
7. Answers will vary with each state.
8. The dotted area represents those states where the population is projected to increase between 5.0% and 9.9%.
9. There are 15 states with the projection between –5.3% and 4.9%, 16 between 5.0% and 9.9%, and 19 between 10.0% and 43.3%. If the District of Columbia is included in the category 10.0% to 43.3%, that number becomes 20, and the total will be 51.
10. The percent of the area of the continental U.S. having an expected range of –5.3% to 4.3% is about 25%. This estimate was made by approximating the area of white regions to that of the rest of the map.
11. The percent of the states in the entire U.S. that has an expected range of 5.0% to 9.9% would be 16/51 = 31.3%.
12. Answers will vary. The most popular answers were yes, the movement seems to be regional; there is an indicated movement to the southwest, warmer climates, and to recreational areas.

The next question should help students discover how they are often influenced by the presentation of the material more than they are by the information itself. Listen to the discussion and interject questions of a similar nature to stimulate the discovery.

13. The maximum projected population loss for the state of Wisconsin would be 4 705 642 × 0.053 = 249 399 people. The maximum projected population gain for the state of Wisconsin would be 4 705 642 × 0.049 = 230 576 people. The remaining discussion questions will vary.
14. The expense of the census makes it prohibitive to be done each year, yet the government needs the information to plan for the future. Using earlier censuses and small samples, the census bureau provides information in years when there is no actual census conducted and before a new census is completed. Making predictions in the year of the actual census and prior to the census also provides an opportunity to examine the accuracy of the methods used for prediction.
15. This question is an attempt to get the group to share their views and come to agreement.
16. Students should recognize how people are influenced as much by the presentation of the data as they are by the actual data.

Be sure students understand they are dealing with percent in working with pyramid graphs. Discuss whether or not it is proper to subtract percents to indicate the difference.

17. In 1890 approximately 13% of the population was between the ages of 30 and 39.
18. Approximately 13% of the population in 1890 was in the age range from 30 to 39 according to this graph. It also indicates approximately 18% will be in that age range in 1990. One conclusion is that the population is getting older, and a smaller percentage of the population is children. In the United States the cost of living is very high, which might be a contributing cause. There is worldwide concern over the growth of the population and the shrinking resources.
19. The distribution of the population of the United States according to age is becoming more equal. This graph indicates an increasing number of older people, perhaps because of better health care, nutrition, and quality of life.
20. The 0–5 age range in 1985 will be the 40–45 age range on the 2030 pyramid graph. The total number of people in that range in 1985 is approximately 17 million.
21. In the year 2030 there will be approximately 21 million persons aged 40–45, an increase of 4 million people. Possible explanations for the difference might be immigration or better counting procedures for the census.
22. The students could create a graph to indicate that the number of elderly will be increasing and the number of younger people to support them will remain constant or diminish. They might possibly identify another problem, so teachers must be alert to their responses.

***Activity 4***

1. The U.S. is smaller because the expected birth rate is less than that of India and China.
2. The probability of a child dying before the age of five in Brazil is between 1 out of 20 and 1 out of 10, or between 0.05 and 0.10.
3. The black region represents the countries in which the probability of a child dying before the age of five is 1 out of 5, or 0.20.
4. There are many contributing factors: famine, poverty, ignorance, climate.
5. With all the wealth and education, concern for others, and worldwide organizations dealing with this problem, there are still a large number of the children of the world dying before age five.
6. There is a chance of 1 in 40 that a child will die before reaching the age of five.

Encourage students to read information from the graphs and use it along with knowledge they already possess to create answers to questions 7–14. The object is not to teach arithmetic methods, although they may need to be reviewed with some students.

7. Each country is drawn according to the small square representing 10 000 foreign-born persons in their population.

8. Six and one-half percent of the population of France was foreign born.
9. According to the legend, between 1 051 288 and 2 478 036 persons living in Poland are foreign born.
10. The fewest foreign-born persons are in Brazil, Mexico, India, Bangladesh, Japan, Egypt, and so on. These countries already have a large native population and few or undeveloped resources.
11. The country could have been a territory of another country. It may have been occupied and the soldiers remained or returned. The country may be very inviting for recreational or occupational reasons, freedom of religion, the type of economy, its stability, and so on.
12. To determine how large to draw a given country, divide the population of foreign-born persons by 10 000. Use this number for the smallest squares in the legend to represent that country's area.
13. The country may be interested in the immigration rate. It may be interested in determining whether it needs to have a second language taught in the country or what the possibilities might be for violent conflict between diverse cultures. The United Nations would be interested in the influence that one particular country would be expected to have over another, and so on.
14. It is much easier to create regions equal to a given area using rectangles and squares than using an irregular shape such as a coastline.

***Activity 5***

1. 75.7% had at least a high school education.
2. 47.7% had not completed high school.
3. Approximately 730 people would not have finished high school.
4. 10.7/21.3 = 50% who started college had finished.
5. 19.9/37.0 = 54% who started college had finished.
6. More people are finishing higher levels of education.

Encourage students to use the histogram to aid in their estimates. The TI-81 graphing calculator would be an excellent way to construct the histogram. Use the trace to help make the estimates. The exact answers are given, but any reasonable answer should be accepted. The answers should also be given in complete sentences.

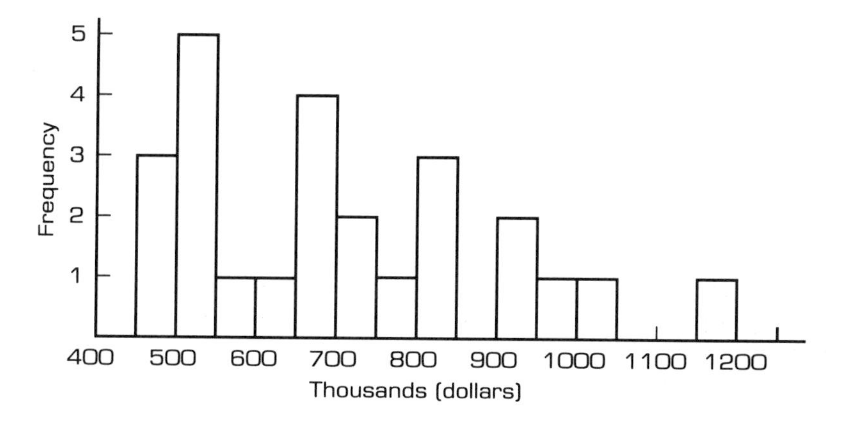


7. Approximately 75% made less than $850 000. ($80%$)
8. The mean winnings was around $700 000. ($702 273)
9. From 50% to 60% of the golfers earned from $550 000 to $950 000.
10. Half of the golfers earned more than $675 000. ($673 433)
11. A typical interval will vary but might be something like this: The typical amount earned by the top golfers was between half a million dollars and a million dollars.
12. Even though there are two in the million-dollar winnings, they are not that far above the others to be outliers.

### *Activity 6*

1. A first impression is that the percentage of drivers with alcohol in their system while driving is generally decreasing. Students could compare the regions and rank them each year for comparison. They should note that the Midwest shows an increase, whereas all other regions show decreases in the same time period. Possible headlines might read, "Midwest Leads the U.S. in Drunken Driving," with the graph to support that position.

### *Activity 7*

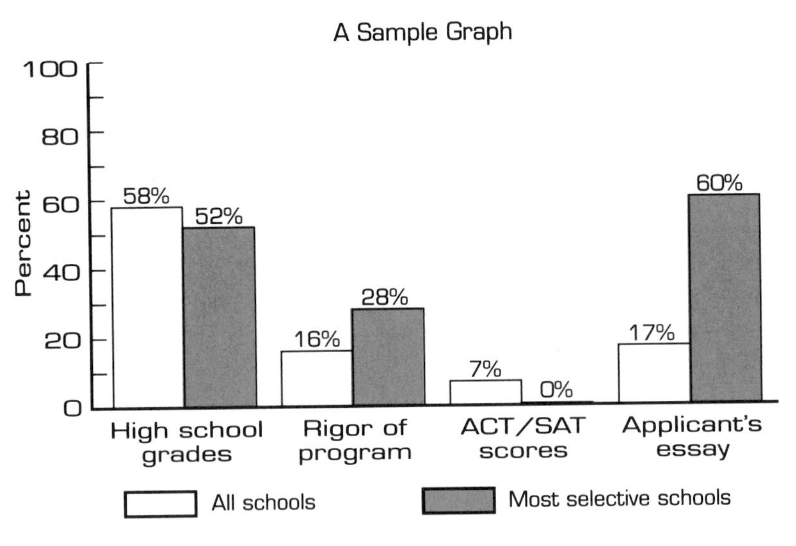


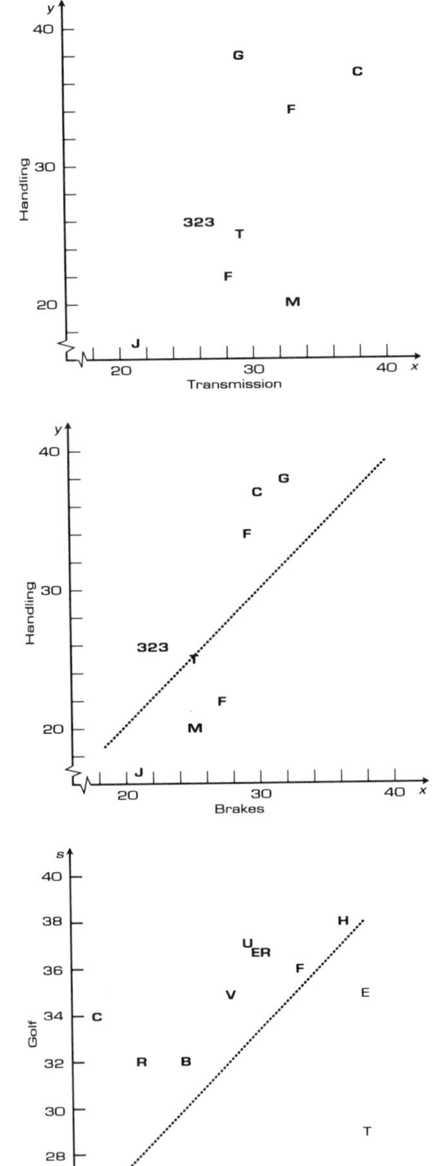


### *Activity 8*

1. a. In general the transmission ratings are higher than the handling ratings. Use a $y = x$ (T = H) line and compare the number of points on each side of the line.
   b. The first graph has a set of points in the upper right: Civic, Fox, and Golf.
   c. Civic rates highest in both categories. The answer can be seen from the graph.
2. a. There is a positive association between brakes and handling. The higher the brake rating, the higher the handling rating. If the brakes are hard to use or grab, the car is harder to steer and control, so the association is reasonable.
   b. Golf is the closest. The angle of the points might make it difficult to determine the closest point. Golf is $\sqrt{68}$ units away from (40, 40); Civic is $\sqrt{109}$.
3. Civic is the best in the three categories, $\sqrt{113}$ from (40, 40, 40).
4. a. Civic rated higher than Golf in the ratings for engine, transmission, style, and style.
   b. The ratings are farthest apart for styling and closest together for handling and fun to drive.
   c. Answers will vary. Be sure students justify their answer in words as well as from the plot.
5. Answers will vary. Have students share their choices and the rationale for them, possibly even asking them to vote at the beginning and again at the end of the sharing to determine whether anyone changed his or her mind after listening to someone's argument.
6. Students might suggest that Honda emphasize its number two place in ranking both on the facts and on the subjective categories or on its acceleration and relatively good mileage.
7. Students might suggest that Subaru advertise its low cost in comparison to the other cars.

### Activity 9

The answers will vary. Encourage students to use such software as Data Insights, Data Models, or Statistics Workshop to make their plots and calculations. Be sure they write their conclusions in language that everyone understands. Help them avoid purely numerical information (such as "the interquartile range is 12") by having them relate the data to their numerical statements.

### Activity 10

1. The data seem to be linear.
2. The slope is approximately −100.
3. The $x$-intercept is about 7.5 and the $y$-intercept is about $750.
4. $y = -100x + 750$
5. You would expect to pay about $350.

Activities such as 11–14 can be used during first-year algebra or the first week of second-year algebra or precalculus to review the properties of linear functions by exploring real data. Discuss domain, range, and slope by using data sets gathered from magazines, almanacs, and other publications. Have students work with partners and then in groups of four to plot and analyze the data. A last assignment might be to bring in one data set of interest. If the examples are not linear, use them later in the semester when teaching techniques to reexpress data.

### Activity 11

1. See graph.
2. Negative
3. Decreases
4. −.150; for every 1-pound increase in weight, the height decreases slightly less than 2/10 of an inch.
5. A 21.5-lb. bike would be able to jump about 10 inches.

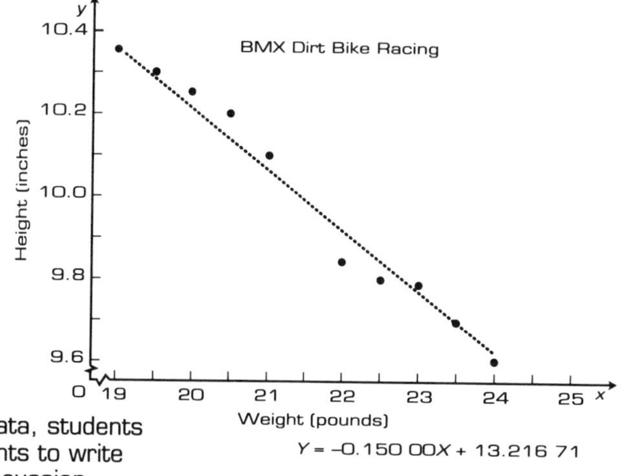


### Activity 12

Students explore different relationships using data from a basketball game. Here students should not only look for an association but also consider the cause-and-effect relation. Does attempting more field goals cause the player to make more? Any association between points scored and fouls could be a function of another variable, playing time. For linear data, students should describe the slope in terms of the data. Encourage students to write complete paragraphs and read some of them to the class for discussion. Median-fit equations are given, but students' answers could vary.

1. Cooper made 2, Scott 1.
2. Answers will vary.

Half of the team scored no fouls regardless of the number of points scored, and the other half has a strong association between fouls and total points scored (probably both are a function of time). An equation would really not describe the data well.

There is a strong positive association between field goals attempted (A) and field goals made (M).

M = 0.64 A −1.03. The slope is about 2/3, which means for every 3 attempts, a player scores 2 field goals.

There is an association between time played ($t$) and points scored ($p$): $p = 0.5t - 3.35$. For every 2 minutes played, a player seems to have a 1-point increase in his score.

There is little association between assists and rebounds, so an equation does not make sense. One player, Johnson, is an outlier—very high in the number of assists and high in rebounds.

### *Activity 13*

This problem about prescription medicine illustrates the importance of slope and reinforces the notion of rate of change. You may instruct students to draw median-fit lines, eyeball lines, or regression lines depending on the background of the class.

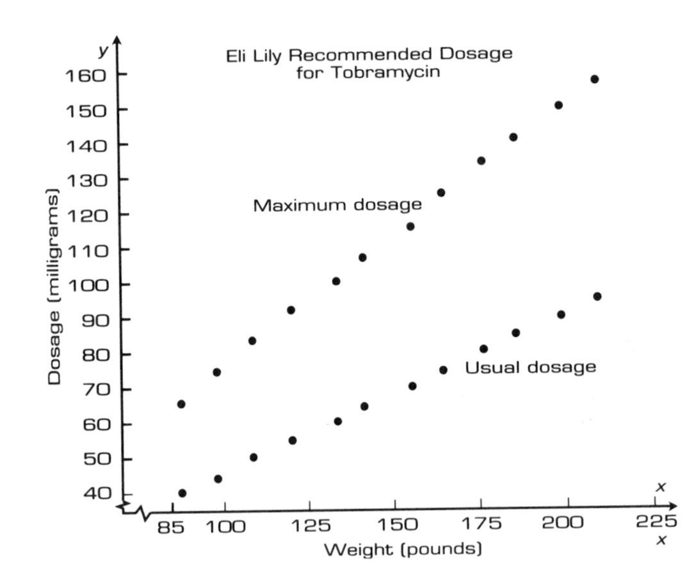


1 and 2. See graph.

3. The slope for usual dosage is about 0.45, and for maximum dosage it is about 0.76. For every pound increase in weight, you can increase the usual dosage by 45%, compared to a 76% increase per pound for maximum dosage.
4. The (weight, usual dosage) equation is approximately $y = 0.46x$. The (weight, maximum dosage) equation is approximately $y = 0.76x - 0.05$.
5. The lines are not parallel because they have different slopes.
6. $y = 1.67x - 0.46$. The ratio of the slopes, the change in maximum dosage to weight to the change in usual dosage to weight (0.76/0.46), is the slope of the new line. (Weight factors out of the ratio.)

### *Activity 14*

Working with Olympic data, students have to determine a scale using minutes and seconds and find the equation of the line. They can interpolate by predicting what might have occurred during the war years as well as look at the danger in extrapolating. The 1988 time was 4:13. The slope is meaningful but might not remain constant as the curve levels off. The domain and range do not include the intercepts. In what year will the winning time for the freestyle be 0 seconds?

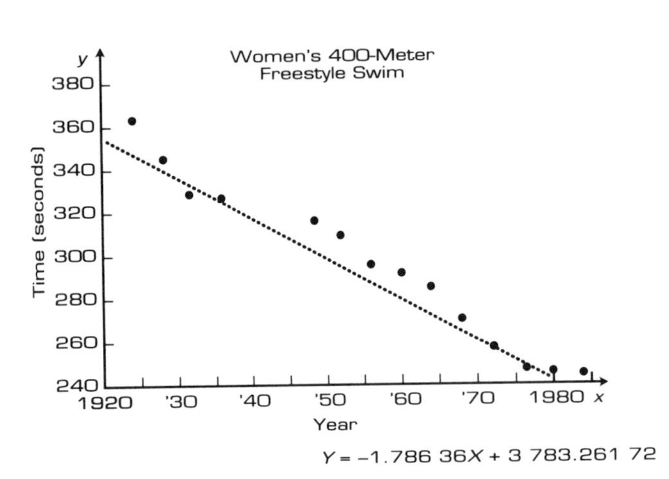


1 and 2. See graph.

3. The slope is $-1.79$ and means that from 1924 to 1984 the winning time for the 400-meter freestyle decreased about 1.79 seconds a year, or about 7 seconds every Olympics.
4. $y = -1.79x + 3783$ (a median-fit line). The time would have been about 5:17 minutes in 1940 and 5:05 minutes in 1944.
5. No, the curve flattens.
6. The winning time in 1988 was 4:03.85 by Janet Evans. The predicted time would be 3:44.48.
7. The regression line is $y = -1.87x + 3934$. Choices might vary; students should compare a graph of both lines.

### *Activity 15*

1. See table 5.2
2. Answers in text.
3. a. (Age, Diameter): $y = 0.185x + 0.554$
   b. (Age, Log Diameter): $y = 0.02x + 0.167$
   c. (Age, Ln Diameter): $y = 0.046x + 0.387$

d. (Log Age, Diameter): $y = 7.80x - 5.287$
e. (Ln Age, Diameter): $y = 3.7x + -5.26$
f. (Ln Age, Ln Diameter): $y = 0.85x + -1.08$
g. (Log Age, Log Diameter): $y = 0.85x - 0.466$

4. The "best" fit seems to be (Ln Age, Diameter).
5. $y = 3.4 \ln x + -5.3$
7. When the tree is 36 years old, the diameter will be about 6.9 inches. At age 50, the diameter will be about 8.0 inches.
8. When the diameter is 8.5 inches, the tree will be about 58 years old.
9. See table 5.3.
10. The root mean squared error for the original data is 1.09, four-tenths of an inch greater error than the logarithmic transformation.
11. $r = 0.92$, which means that there is a strong positive association between the height and the logarithm of the age of a tree.
12. If the correlation were 0.3, the plot would have little pattern. Some of the young trees would have large diameters, and some of them would have small diameters.

***Activity 16***

1.

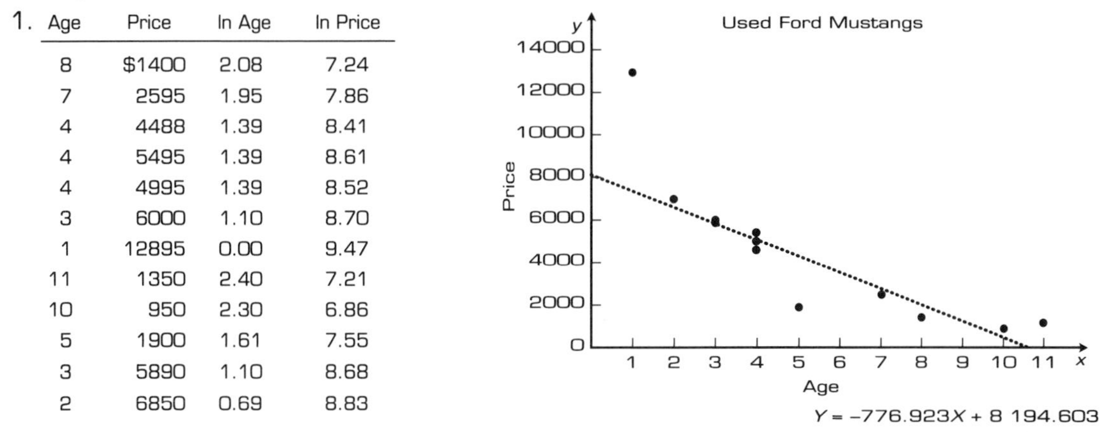

| Age | Price | ln Age | ln Price |
|---|---|---|---|
| 8 | $1400 | 2.08 | 7.24 |
| 7 | 2595 | 1.95 | 7.86 |
| 4 | 4488 | 1.39 | 8.41 |
| 4 | 5495 | 1.39 | 8.61 |
| 4 | 4995 | 1.39 | 8.52 |
| 3 | 6000 | 1.10 | 8.70 |
| 1 | 12895 | 0.00 | 9.47 |
| 11 | 1350 | 2.40 | 7.21 |
| 10 | 950 | 2.30 | 6.86 |
| 5 | 1900 | 1.61 | 7.55 |
| 3 | 5890 | 1.10 | 8.68 |
| 2 | 6850 | 0.69 | 8.83 |



2. The data are not linear. The points are below the line in the center and above the line at the ends. If the data were linear, the slope would be the rate of change of the price per year of age for the cars, or depreciation. For this line, $y = -777x + 8194$, and the depreciation would be a loss of $777 a year in the value of the car. The *y*-intercept would represent the price of the car when it was 0 years old, or a new car, and the *x*-intercept would represent the age at which the car was worth no money. A new Mustang should cost approximately $8194, and the car would be worthless when it was about 10.5 years old. Depreciation is not a constant, however, for cars depreciate faster when they are new; the model should not be linear.
3. The best transformation seems to be ($\ln x$, $y$). The equation of the median-fit line would be $y = -3900 \ln x + 9995$.

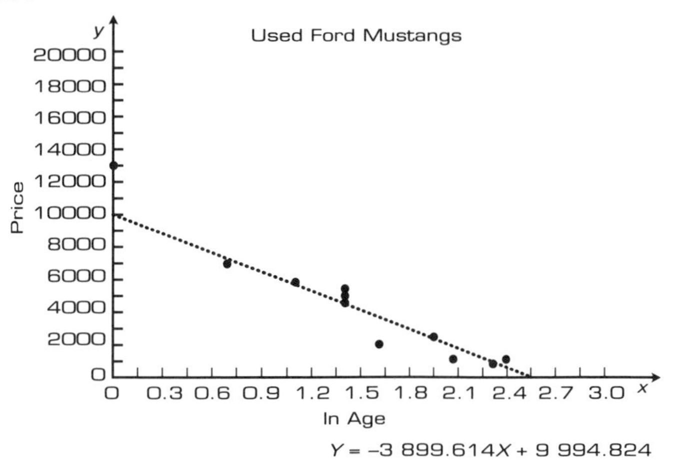

4. The correlation coefficient is 0.96. This indicates a very strong association between price and the logarithm of the age of the car; ln age is a good predictor for the price of a used car. This is also clear from the graph.
5. A used six-year-old Mustang would sell for about \$3022.
6. For \$4000, you could buy a four-year-old car.

***Activity 17***

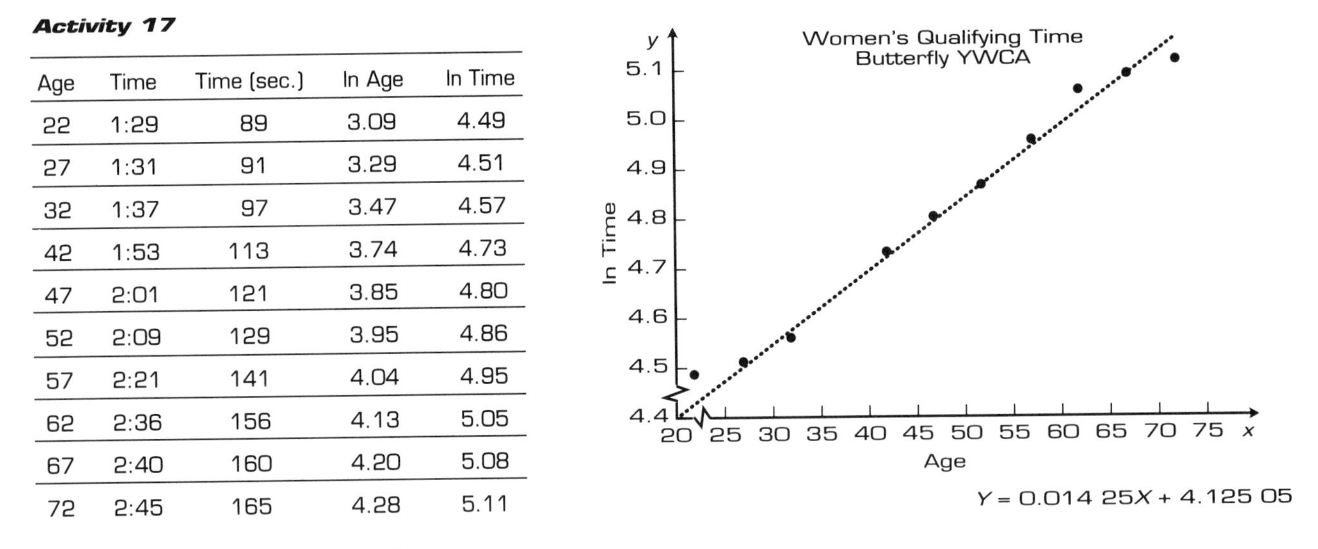

| Age | Time | Time (sec.) | ln Age | ln Time |
|---|---|---|---|---|
| 22 | 1:29 | 89 | 3.09 | 4.49 |
| 27 | 1:31 | 91 | 3.29 | 4.51 |
| 32 | 1:37 | 97 | 3.47 | 4.57 |
| 42 | 1:53 | 113 | 3.74 | 4.73 |
| 47 | 2:01 | 121 | 3.85 | 4.80 |
| 52 | 2:09 | 129 | 3.95 | 4.86 |
| 57 | 2:21 | 141 | 4.04 | 4.95 |
| 62 | 2:36 | 156 | 4.13 | 5.05 |
| 67 | 2:40 | 160 | 4.20 | 5.08 |
| 72 | 2:45 | 165 | 4.28 | 5.11 |



$Y = 0.014\ 25X + 4.125\ 05$

1. See graph. The data are not linear.
2. The transformation is an exponential transformation. The equation of the median-fit line is $\ln y = 0.014x + 4.12$. This can be rewritten as $y = e^{0.014x + 4.12} = e^{0.014x} e^{4.12} = 62.1 e^{0.014x} = 61.8(1.01^x)$.
3. The predicted time is 104 seconds, which is 1 second from the actual qualifying time.
4. The mean squared error is nearly 0, which means the line is a very good fit for the data.
5. The transformations are as follows: linear: $y = 46 + 1.7x$, $r = 0.990$; logarithmic: $y = 69 \ln x - 137$, $r = 0.96$; exponential: $y = 64\ (1.01)^x$, $r = 0.994$; power: $y = 14x^{0.57}$, $r = 0.98$.
6. The median-fit line is based on medians, which might be different from the mean on which the regression line produced by the calculator is based. Over the small domain and set of data points produced, there is not that much difference in some of the models.

***Suggestions on using technology:***

Use the following steps to calculate the root mean squared error on the Casio 7500 or 7000 graphing calculators. Set the mode for writing programs, Mode 2. Select a program number # and type ? –M:?–N:m*ln M+b–N. To run the program, set Mode 1 and type Prog # Execute and a question mark will appear on the screen. Input the first *x* (age) and execute. A second question mark will appear; input the corresponding *y* (diameter) and execute. The difference will appear as the result. If the calculator is in LR1 (Mode –), this difference can be recorded or entered directly as a data point. When all the differences have been entered, $\Sigma x^2$ will give the sum of the squares. Divide by the number of data points to find the mean squared error, and take the square root to find the average distance or standard deviation of the points from the line.

To find and plot the residuals and calculate the root mean error on the TI-81, see *Exploring Statistics on the TI-81* (Burrill and Hopfensperger 1991).

A simple BASIC program will also generate the calculations and the error for this problem. The following program will run on an Apple II family of computers. (NOTE: For this problem, type in $n = 27$, $m = 3.4$, and $b = -5.35$. To run the program, enter D = Domain and R = Range for the given data points.)

```
10 L = 0
20 FOR x = 1 to n
30 INPUT D,R
40 y = m*log (D) + b
50 S = INT (1000*Y+.5)/1000
60 M = INT (1000*(R–S)+.5)/1000
70 N = INT (1000*M²+.5)/1000
80 L = L+N
90 PRINT D,R,S,M,N,L
100 NEXT X
```

## REFERENCES


Bryan, Elizabeth H. "Exploring Data with Box Plots." *Mathematics Teacher* 81 (November 1988): 658–63.

Burrill, Gail F. "Statistics and Probability." *Mathematics Teacher* 83 (February 1990): 113–18.

Burrill, Gail, and Patrick Hopfensperger. *Exploring Statistics with the TI-81*. Menlo Park, Calif.: Addison-Wesley Publishing Co., 1992.

de Lange, Jan, and M. Kindt. *Grafische Verwerking*. Educaboek, The Netherlands: Culemborg, 1985.

de Lange, Jan, and H. van der Kooij. *Statistiek*. Utrecht: Utrecht University, OW & OC, 1989.

de Lange, Jan, and Heleen B. Verhage. *Data Visualizations*. Experimental version. Utrecht: Utrecht University, OW & OC, 1989.

Freedman, David, Robert Pisani, and Roger Purves. *Statistics*. New York: W. W. Norton & Co., 1980.

Gnanadesikan, Mrudulla, Richard Scheaffer, and James Swift. *The Art and Technique of Simulation*. Palo Alto, Calif.: Dale Seymour Publications, 1987.

Hoffman, Mark, ed. *The World Almanac and Book of Facts, 1988*. New York: Random House, 1988.

______. *The World Almanac and Book of Facts, 1990*. New York: Random House, 1990.

Landwehr, James, and Ann E. Watkins. *Exploring Data*. Palo Alto, Calif.: Dale Seymour Publications, 1986.

Landwehr, James, James Swift, and Ann E. Watkins. *Exploring Surveys and Information from Samples*. Palo Alto, Calif.: Dale Seymour Publications, 1987.

Lappan, Glenda, William Fitzgerald, Elizabeth Phillips, Janet Shroyer, and Mary Jean Winter. *Probability*. Middle School Mathematics Project. Menlo Park, Calif.: Addison-Wesley Publishing Co., 1986.

Link, Richard F. "Election Night on Television." In *Statistics: A Guide to the Unknown*, 2d ed., pp. 178–86. San Francisco: Holden-Day, 1978.

National Council of Teachers of Mathematics. *Curriculum and Evaluation Standards for School Mathematics*. Reston Va.: The Council, 1989.

North Carolina School of Science and Mathematics. *Data Analysis*. New Topics for Secondary School Mathematics. Reston, Va.: National Council of Teachers of Mathematics, 1988.

______. *Geometric Probability*. New Topics for Secondary School Mathematics. Reston, Va.: National Council of Teachers of Mathematics, 1989.

Shulte, Albert P., and Jim Swift. "Data Fitting without Formulas." *Mathematics Teacher* 79 (April 1986): 264–71.

Steen, Lynn A., ed. *For All Practical Purposes, Introduction to Contemporary Mathematics*. New York: W. H. Freeman & Co., 1988.

Travers, Kenneth J., William Stout, James Swift, and Joan Sextro. *Using Statistics*. Menlo Park, Calif.: Addison-Wesley Publishing Co., 1985.

Tufte, E. R. *The Visual Display of Quantitative Information*. Cheshire, Colo.: Graphic Press, 1983.

Woodward, Ernest, and Jim R. Ridenhour. "An Interesting Probability Problem." *Mathematics Teacher* 75 (December 1982): 765–68.

◆ ◆ ◆ ◆ ◆ ◆ ◆ ◆

## BIBLIOGRAPHY FOR STATISTICS

### Books

Burrill, Gail F., and Patrick Hopfensperger. *Exploring Statistics on the TI-81*. Menlo Park, Calif.: Addison-Wesley Publishing Co., 1991.

This text explains step by step how to use the graphing calculator to explore data, estimate probabilities by simulations, and develop sampling distributions. The unique features of the calculator are integrated in the presentation of statistical techniques, and exercises stress interaction among the data, the numerical statistics, and the graphical representations.

"Data Analysis." February 1990 *Mathematics Teacher*. Reston, Va.: National Council of Teachers of Mathematics, 1990.

This minifocus issue on statistics contains articles on using box plots, implementing the curriculum standards for statistics and probability, the use and misuse of statistics, and building the concept of inference by a simple version of the randomization test. Many ideas for classroom use are present as well as an article providing motivation for using the new data analysis techniques.

de Lange, Jan, and Heleen Verhage. *Data Visualizations.* Utrecht: Utrecht University, 1990.

In this booklet on using information from graphs, tables, and charts to solve problems, the focus is on developing a critical attitude about data. Students are encouraged to look at data from a variety of perspectives and to use the information they obtain to make decisions. The problems are embedded in a context, and the students are expected to read in order to understand and solve the problems.

Freedman, David, Robert Pisani, and Roger Purves. *Statistics.* New York: W. W. Norton & Co., 1980.

This book, written in everyday language, presents standard calculations but emphasizes learning to think statistically. Different statistical models are analyzed with examples chosen from a variety of disciplines to show the use of statistics as a tool. The reading level is difficult for most high school students, but the book is an excellent teacher reference.

Gnanadesikan, Mrudulla, Richard Scheaffer, and James Swift. *The Art and Technique of Simulation.* Palo Alto, Calif.: Dale Seymour Publications, 1987.

Part of the Quantitative Literacy Series, this booklet introduces simulation to explore the behavior of complex situations, estimate probabilities, summarize data, and predict results. An appendix contains suggested computer programs, but the primary emphasis is on a hands-on approach without complicated formulas or algebraic manipulations. The problems are realistic and furnish good examples of real-life applications of probability and statistics.

Huff, Darrell. *How to Lie with Statistics.* New York: W. W. Norton & Co., 1959.

This is the classic book on the misuse of statistics. Although the examples are old, the message has remained unchanged. The book covers the topic very well and can be easily read by students as well as serve as an excellent teacher resource.

Landwehr, James, and Ann E. Watkins. *Exploring Data.* Palo Alto, Calif.: Dale Seymour Publications, 1986.

The first of the Quantitative Literacy Series, the book emphasizes data analysis and stresses the importance of organizing and displaying data so they reveal patterns and trends. A variety of new and easy-to-use graphical methods to display data are introduced. The examples are from topics of interest to students, and the text is a must for anyone interested in teaching statistics at the secondary school level.

Landwehr, James, James Swift, and Ann E. Watkins. *Exploring Surveys and Information from Samples.* Palo Alto, Calif.: Dale Seymour Publications, 1987.

Focusing on the use of simulation to generate samples, this last book in the Quantitative Literacy Series introduces students to the concept of confidence intervals in a unique and graphic way. Sampling strategies are discussed as well as the design and use of surveys and the role of polling organizations such as the Gallup Poll. The book is an excellent introduction to statistical inference.

Moore, David S. *Statistics: Concepts and Controversies.* New York: W. H. Freeman & Co., 1988.

This is a book for students and citizens on statistical ideas and their importance in public policy and in human sciences. The intent is to teach verbally rather than symbolically for people interested in ideas rather than techniques. At the end of each chapter there are good examples and many exercises that explore concepts such as correlation in lay terms. This is another must as a resource for anyone interested in teaching statistics.

______. "Uncertainty." In *On the Shoulders of Giants: New Approaches to Numeracy,* edited by Lynn A. Steen. Washington, D. C.: National Academy Press, 1990.

This chapter deals with uncertainty (data and chance), one of five major strands suggested as organizing themes for curricula in the next century. The chapter is replete with powerful illustrations of data representation, analysis, and inference supporting the kinds of quantitative reasoning with which all students should be familiar.

Moore, David S., and George McCabe. *Introduction to the Practice of Statistics.* New York: W. H. Freeman & Co., 1989.

An elementary introduction to college statistics, this book can be used with the video series Against All Odds: Inside Statistics. The text focuses on data and statistical reasoning and uses many modern graphical techniques. Suggested exercises for use with software packages are an integral part of the book, which is designed for students with only an algebra background. This is an excellent resource.

National Council of Teachers of Mathematics. *Teaching Statistics and Probability.* 1981 Yearbook, edited by James R. Smart. Reston, Va.: The Council, 1981.

A collection of articles and essays on teaching, the book includes articles on curriculum issues and applications of statistics outside the classroom. The material discusses how people and mathematics cope with uncertainty and provides ideas for classroom models, strategies, activities, and projects.

North Carolina School of Science and Mathematics. *Data Analysis.* New Topics for Secondary School Mathematics. Reston, Va.: National Council of Teachers of Mathematics, 1988.

Part of the NCTM series New Topics for Secondary School Mathematics, the materials are part of a course designed to prepare high school students who have completed two years of algebra for a variety of college mathematics courses. Applications-oriented and based on student investigations and experiments, the booklet begins with simple data analysis techniques. A primary emphasis is on reexpressing data using mathematical models for nonlinear functions. Computer software for the IBM PC and a guide are available with the materials.

Tanur, Judith M., Frederick Mosteller, William H. Kruskal, E. Lehmann, Richard F. Link, Richard S. Pieters, and Gerald R. Rising. *Statistics: A Guide to the Unknown,* 3d ed. Belmont, Calif.: Wadsworth, 1989.

An early product from the ASA/NCTM Joint Committee on Statistics, the third edition of this book is an updated series of essays on statistics classified in four broad areas of application: the biological, social, physical, and political world. Written for readers without any special knowledge, the aim of the essays is to show how important problems in those fields were solved using probability and statistics. The examples are excellent and can serve as focal points for student projects or research.

Travers, Kenneth J., William Stout, James Swift, and Joan Sextro. *Using Statistics.* Menlo Park, Calif.: Addison-Wesley Publishing Co., 1985.

Intended for students who have completed one year of high school algebra, the text covers descriptive statistics, introduces students to probability, and furnishes an elementary introduction to hypothesis testing. Many of the exercises are laboratory-based and provide students with an experience base to use in making statistical decisions.

Tufte, Edward R. *The Visual Display of Quantitative Information.* Cheshire, Colo.: Graphics Press, 1983.

Called a "tour de force" by a reviewer, this book studies the communication of information through the simultaneous presentation of words, numbers, and pictures. The first part is a study of the history of graphical design, and the second part provides a language for discussing graphics and a set of principles to use in this design. There are many wonderful pictures and computer designs that can furnish a rich source for class discussion.

### *Software*

Cricket Graph. San Jose, Calif.: Computer Associates, 1986.

Available for the Macintosh, IBM, and compatibles, this software works with MicroSoft Windows. Using their own data, students can experiment with a variety of fitted lines and curves. There are several options for each screen that provide different ways to explore the data.

Data Insights. Pleasantville, N. Y.: Sunburst Communications, 1989.

Available for Apple IIe, IBM, and Tandy computers, this software enables students to work easily with large sets of data. With little instruction they can display data in a variety of plots and calculate statistics. Every screen can be printed and used for analysis outside the classroom. Using this software as a tool, students have time to experiment with data and can concentrate on the analysis necessary to solve problems. A manual of student materials is a part of the package.

Data Models. Pleasantville, N. Y.: Sunburst Communications, 1991.

Available for the Macintosh, this easy-to-use software enables students to enter paired data and fit lines and curves. It also provides equations, correlation coefficients, residuals, and their graphs. Students can investigate data transformations in both tabular and graphic form.

Statistics Workshop. Designed by BBN Laboratories, Pleasantville, N. Y.: Sunburst Communications, 1991.

Available for Apple IIe, Macintosh, IBM, and Tandy computers, this software allows students to explore histograms, means, box plots, scatter plots, and—a unique feature—to investigate categorical data. It provides "stretchy" histograms to allow students to manipulate distributions and explore what happens to the descriptive statistics. Students are able to move fitted lines visually through the graphs of paired data to determine relationships.

### *Videos*

Moore, David, ed. Against All Odds: Inside Statistics. Annenberg/CPB Project Series. Arlington, Mass: Consortium for Mathematics and Its Applications (COMAP), 1989.

This video series is a 26-program telecourse on statistics and its applications. The course parallels *An Introduction to the Practice of Statistics* and offers a visual introduction to modern statistics. Excerpts can be valuable supplements to classroom instruction. A Study Guide with an overview of the content, exercises, and self-test questions is available for students. There is an instructor's manual, which has advice on implementing a telecourse.